T0330532

Robust Quality

Continuous Improvement Series

Series Editors
Elizabeth A. Cudney and Tina Kanti Agustiady

Published Titles

Affordability: Integrating Value, Customer, and Cost for Continuous Improvement
Paul Walter Odomirok, Sr.

Continuous Improvement, Probability, and Statistics: Using Creative Hands-On Techniques
William Hooper

Design for Six Sigma: A Practical Approach through Innovation
Elizabeth A. Cudney and Tina Kanti Agustiady

Statistical Process Control: A Pragmatic Approach
Stephen Mundwiller

Transforming Organizations: One Process at a Time
Kathryn A. LeRoy

Forthcoming Titles

Robust Quality: Powerful Integration of Data Science and Process Engineering
Rajesh Jugulum

Building a Sustainable Lean Culture: An Implementation Guide
Tina Agustiady and Elizabeth A. Cudney

Robust Quality

Powerful Integration of Data Science and Process Engineering

Rajesh Jugulum

CRC Press is an imprint of the
Taylor & Francis Group, an **informa** business

CRC Press
Taylor & Francis Group
6000 Broken Sound Parkway NW, Suite 300
Boca Raton, FL 33487-2742

Printed on acid-free paper

International Standard Book Number-13: 978-1-4987-8165-7 (Hardback)

Visit the Taylor & Francis Web site at
http://www.taylorandfrancis.com

and the CRC Press Web site at
http://www.crcpress.com

Contents

Foreword

The world of data and technology and the world of process improvement and quality have long existed in parallel universes with no bridge or wormhole between them. Business processes are improved without any consideration of information technology—the most powerful enabler of better process performance. Technology is applied to support process tasks that shouldn't exist in the first place. Metrics and analytics abound throughout the organization, but don't really relate to key business processes. Many organizations have a focus on product and process quality, but no orientation to data quality at all.

There is a reason for why these ideas have remained within silos. The forefathers of quality and process improvement didn't have information technology at hand as a tool to improve their methods. Technology, data, and analytics were viewed as an adjunct to business, rather than the core of it. Process improvement and quality approaches were focused only on production activities—the core of an economy then devoted to manufacturing.

This book, by Rajesh Jugulum, is an antidote to this regrettable state of affairs. It forges a strong connection among the concepts of data science, analytics, and process engineering. In the digital era, it is impossible to create the needed levels of performance improvement in organizations without harnessing data and technology as levers for change. This book presents these tools in the context of quality and process management and also ties them to business strategy and operations.

The concepts are derived from the synthesis of analytics, quality, process, and data management, but they apply to a broader context. They address the improvement and management of quality in products, strategies, and even the broader society in which organizations operate. They also provide a useful perspective on each of the underlying activities. Analytics, for example, is typically viewed in a context of data and algorithms alone. However, they are also a process that can be analyzed, measured, and improved. The book's content on *metrics management* and *analytics quality* provides a needed guide to the ongoing management of these important resources. The topics even extend to the current focus on machine learning and artificial intelligence.

These concepts may seem a bit abstract as they are discussed. However, Jugulum has provided a number of examples in industries like health care, finance, and manufacturing to flesh out the details and context of his approach. Ultimately, organizations will need to provide their own context, trying out the models and approaches in the book and noting how they change behavior.

Jugulum is well-suited due to his background to tie these previously disparate concepts together. As a practitioner, he's worked in analytics, technology strategy, process management, and data management at several large financial services and health care firms. In addition to his day jobs, he's developed a set of analytical software tools that measure improvements in a business process using the concepts behind Hoshin Kanri, a Japanese approach to quality and process deployment.

He's written books on relating Lean Six Sigma approaches to product and process design and on the importance of having a structured approach for improving data quality. It is rare to find this combination of theory and practice in one individual.

We clearly live in an age where information systems, data, and analytics have become the primary production capabilities for many organizations. Products and services incorporate them as essential features. Operational processes can't be carried out without them.

In this business environment, it makes little sense to discuss quality and processes as if they only applied to traditional manufacturing. This book is an important step in the extension of quality and process improvement concepts to the fields of data and analytics. Today, it is a novel approach; in the near future, we will wonder how we ever proceeded otherwise.

Thomas H. Davenport
Distinguished Professor of IT and Management, Babson College
Fellow, MIT Initiative on the Digital Economy
Author of *Process Innovation, Competing on Analytics,* and *Only Humans Need Apply*

Preface

The term *quality* is usually used in the context of manufacturing a product or service as a measure of performance. With a rapid growth in data with use of technology, mobile devices, social media, and so on, many companies have started to consider data as critical asset and to establish a dedicated data management function to manage and govern data-related activities to ensure the quality of data is good and that they are fit for the intended use. Along with their growth in the data discipline, organizations have also started using analytics quite significantly to make sound business decisions and, therefore, the quality of analytics is of equal importance. Because of these changes in day-to-day activities, especially through data, the term quality should be expanded as a measure of process, data, and analytics. Quality should be measured with a holistic approach.

We use the term *robust quality* to measure quality holistically. Therefore, the subject of robust quality should include all concepts/tools/techniques in process engineering, and data science. Often times, companies fail to recognize the relationship between these two disciplines. Mostly, they operate in silos. The proposed approaches in this book will help to establish these relationships and quickly solve business problems more accurately. They also focus on aligning quality (data and process) strategy with corporate strategy and provide a means for execution. These methods can help change industry culture and assist organizations with becoming more competitive in the marketplace. With strong leadership and the implementation of a holistic robust quality method, the quality journey will be more successful and yield positive results. For improving products/services, the data quality approach can be integrated with Lean Six Sigma and Dr. Genichi Taguchi's quality engineering philosophies. For improving analytics quality, the data quality approach is integrated with the general process of analytics execution and purposeful analytics methods.

The methods provided in this book will guarantee improvements and stable, predictable, and capable operations. The case studies presented will empower users to be able to apply these methods in different real-life situations while understanding the methodology. I believe that the application of the methods provided in this book will help users achieve robust quality with respect to product development activities, services offered, and analytical-based decision-making.

Dr. Taguchi, whom I consider to be the greatest of all time in the field of quality engineering, always used to say that quality has an inverse relation to loss to society. If data quality and/or process quality are not satisfactory, it will result in bad products/services. If data quality and/or analytics quality are not satisfactory, it will also result in poor decisions. All these things will add up and there will be a huge loss to society. The methods provided in this book, to a large extent, are intended to minimize the loss to society. To successfully apply these methods, we need to change the process of thinking and act differently with data and analytics. Data and analytics-based thinking not only helps in making sound business decisions but also plays a major role in day-to-day decision-making activities. I will conclude this section with a related quote from H.G. Wells.

> "Statistical thinking will one day be as necessary for efficient citizenship as the ability to read and write."

I venture to modify this quote so that it aptly suits the data and analytics-driven world as:

> "Data and analytical thinking are as necessary for efficient citizenship as the ability to read and write."

Rajesh Jugulum
May 2018

Acknowledgments

Book-writing is always a challenge and a great experience, as it can involve efforts to summarize the building of new ideas, the development of a framework for their execution in the real world. It may also involve the use of concepts/philosophies of several distinguished individuals and the gathering of inputs from many talented people. First, I would like to thank the late Dr. Genichi Taguchi for his outstanding thought leadership in the area of quality engineering. His philosophy remains key in the development of a robust quality approach that integrates data science and process engineering.

I am very grateful to Professor Tom Davenport for his support and encouragement to this effort by writing the Foreword. I consider myself fortunate to receive support from a well-known and well-respected person like Professor Davenport.

I would also like to thank Professor Nam P. Suh for developing axiomatic design theory, which is benefiting to society in many ways. Chapter 3 of this book presents a description of the building of data and process strategy using axiomatic design principles. Thanks are also due to Brian Bramson, Bob Granese, Chuan Shi, Chris Heien, Raji Ramachandran, Ian Joyce, Jagmeet Singh, Don Gray, and John Talburt for their involvement in putting together a data quality approach and conducting a case study that is presented in Chapter 6. Thanks are also due to Laura Sebastian-Coleman, Chris Heien, Raj Vadlamudi, and Michael Monocchia for their efforts/help in conducting case studies that are also presented in Chapter 6.

My thanks are always due to the late Professor K. Narayana Reddy, Professor A. K. Choudhury, Professor B. K. Pal, Mr. R. C. Sarangi, and Professor Ken Chelst for their help and guidance in my activities. I am also grateful to Randy Bean, Phil Samuel, Leandro DalleMule, Gabriele Arcidiacono, Elizabeth Cudney, Tirthankar Dasgupta, Javid Shaik, and Sampangi Raman for their support and help during this activity.

I wish to express my gratitude to Cigna, especially to CIMA and the ethics office for allowing me to publish the book while I am employed at Cigna. Thanks are due to Lisa Bonner and Karen Olenski of CIMA for providing the necessary help in getting the required approvals and for supporting this activity. I am also thankful for the involvement of Kanri, Inc. while some aspects of the *purposeful analytics* section of Chapter 5 were being developed.

I am thankful to John Wiley & Sons and Springer for allowing me to use parts of my previous publications. I am very grateful to CRC Press for giving me an opportunity to publish this book. I am particularly thankful to Cindy Renee Carelli, Executive Editor, for her help and support and for being flexible in accommodating many requests from me. Thanks are also due to Joanne Hakim of Lumina Datamatics for her help and cooperation during this effort.

I am also grateful to my father-in-law, Mr. Shripati Koimattur, for carefully reading the manuscript and providing valuable suggestions. Finally, I would like to thank my mother and my family for their understanding and support throughout this effort.

Author

Rajesh Jugulum, PhD, is the Informatics Director at Cigna. Prior to joining Cigna, he held executive positions in the areas of process engineering and data science at Citi Group and Bank of America. Rajesh completed his PhD under the guidance of Dr. Genichi Taguchi. Before joining the financial industry, Rajesh was at Massachusetts Institute of Technology where he was involved in research and teaching. He currently teaches at Northeastern University in Boston. Rajesh is the author/co-author of several papers and four books including books on data quality and design for Six Sigma. Rajesh is an American Society for Quality (ASQ) Fellow and his other honors include ASQ's Feigenbaum medal and International Technology Institute's Rockwell medal. Rajesh has delivered talks as the keynote speaker at several conferences, symposiums, and events related to data analytics and process engineering. He has also delivered lectures in several universities/companies across the globe and participated as a judge in data-related competitions.

1 The Importance of Data Quality and Process Quality

1.1 INTRODUCTION

As a result of the data revolution, many organizations have begun to view data as important and critical asset; that is, with a level of importance equal to those of other resources such as people, capital, raw materials, and infrastructure. This has driven the need for dedicated data management programs. However, beyond ensuring data are fit for their intended business purposes, organizations must also focus on the creation of shareholder value through data-related activities. To achieve this, organizations should focus on developing a data and analytics strategy with characteristics such as speed, accuracy, and precision of data, as well as analytics management processes to help differentiate the organization from competitors. Such a strategy should also be aligned with the corporate strategy so that data and analytics requirements can be effectively prioritized.

In order to ensure data are fit for the purpose, we must have high-quality levels of data. Data quality is related to process quality (or Six Sigma quality) in two ways: (1) when working on Six Sigma initiatives, ensuring data quality is important to ensure high-quality performance levels for products/systems; and (2) data quality is also dependent upon certain processes and, as we make improvements to these processes, data quality needs to be improved as well. The overall goal of this book is to provide an integrated approach by combining data science and process engineering to achieve robust quality.

In this introductory chapter, we discuss the importance of the concepts of data quality and process quality and why the integration of both is required to address quality holistically. This chapter will also discuss data as being important asset to any organization like how people or infrastructure are and how data management programs are being built to effectively treat the data. Discussion will also focus on the impact of poor data quality and how it contributes to societal loss using the theory of Taguchi's loss function, as well as the impact of data quality on process improvements and why the integration of data quality and process quality is necessary.

1.2 IMPORTANCE OF DATA QUALITY

Data capability is increasingly becoming critically important in this information-driven world. It is believed by many that this capability should be viewed in the same positive manner as other assets of an organization, such as people, infrastructure, and raw materials. This thought process has driven the need to manage data across organizations in a disciplined fashion that will help users to derive meaningful insights that will eventually drive business excellence.

Implications of Data Quality

Dr. Genichi Taguchi, a Japanese expert in the field of quality engineering (QE) (that the author was fortunate to work with), emphasized the importance of having good quality products to minimize overall loss. Taguchi (1987) established a relationship between poor quality and overall loss by using a quadratic loss function (QLF) approach. Quality loss function describes the loss that a system produces from an adjustable characteristic. According to the QLF concept, the loss increases if the characteristic y (such as speed or strength) is away from the target value (m)—meaning, there is a loss associated if the quality characteristic moves away from the target. Taguchi's philosophy terms this loss as a loss to society, and someone has to pay for this loss. Here, that "someone" is a part of society, whether it be customers, organizations, or the government. These losses will have adverse effects, resulting in system breakdowns, accidents, unhappy customers, company bankruptcies, and so on. Figure 1.1 shows how the loss increases if the characteristic deviates from the target (on either side) by Δ_0, and is given by $L(y)$. When y is equal to m, the target value, then the loss is zero or at the minimum. The equation for the loss function can be represented as follows:

$$L(y) = k(y - m)^2$$

where k is a factor that is expressed in dollars, based on different types of costs such as direct costs, indirect costs, warranty costs, reputational costs, monetary loss due to lost customers, and costs associated with rework and rejection.

It is important to note that the QLF is usually not exactly symmetrical and, as most cost calculations are based on estimations or predictions, a close approximate function is quite adequate.

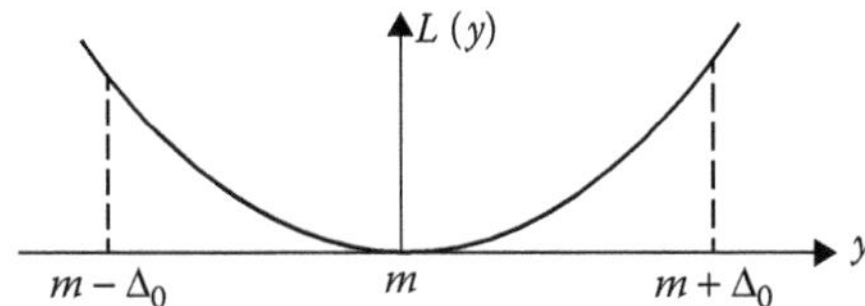

FIGURE 1.1 Quadratic loss function (QLF).

In the data quality world, the concept of the loss function is a very useful thing when we are dealing with the quality of data elements such as customer social security numbers, customer addresses, and account balances. From the list of data elements, critical data elements (CDEs) are selected based on certain criteria. The data quality of these CDEs is typically measured in terms of percentages. These percentages are based on individual dimensional scores that are based on data accuracy, data completeness, data conformity, and data validity.

If the data quality levels associated with these CDEs are not on or close to the target values, then there is a high probability of making incorrect decisions, which could lead to adverse effects on organizations. Since the data quality levels are of *the higher, the better* type (i.e., a higher percentage is better), only half of the QLF is applicable when measuring loss due to poor data quality. The loss function in the context of data quality is shown in Figure 1.2. Through this figure, one can see how the overall loss increases if the data quality level of a CDE is away from the target m. Sometimes, in a data quality context, the target values are also referred to as the business specifications, or thresholds.

As we can see in Figure 1.2, the loss will be at a minimum when y reaches the target level m. This loss will remain at this level even if the quality levels improve beyond m. So, sometimes, it may not be necessary to improve the CDE quality levels beyond m.

Poor data quality may incur losses in several forms. They include (English, 2009) the impacts of the denial of a scholarship to a student for college and the placement of inaccurate labels on products.

Taguchi (1987) classifies the effects of poor quality into two categories: (1) losses caused due to functional variability of the products and processes, and (2) losses caused due to harmful side effects. Figure 1.3 shows how all of these costs—and one can imagine how they might add up—cause overall loss to society.

The importance of ensuring high-quality data was emphasized by famous statisticians long before the data field experienced massive growth. A famous British statistician, Ronald A. Fisher, mentioned that the first task of a statistician/analyst is to carry out the cross-examination of the data so that a meaningful analysis of the data and an interpretation of the results can be done. Calyampudi R. Rao, a world-renowned Indian statistician, provided a checklist (Rao, 1997) for cross-examination of the data, in which emphasis was primarily given to the data quality and analysis of the measurement system that we use for data collection.

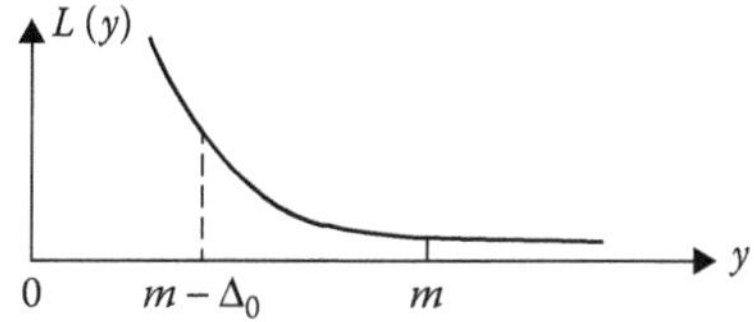

FIGURE 1.2 Loss function for data quality levels *(higher the better)* characteristic.

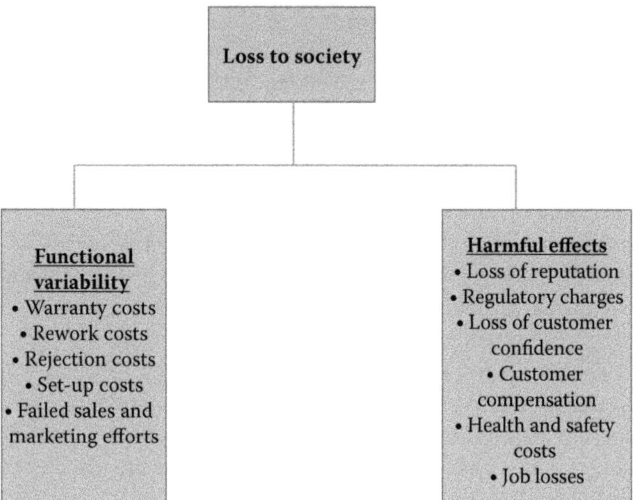

FIGURE 1.3 Loss to society sources.

Data Management Function

As mentioned earlier, several organizations have started to implement dedicated data management functions that are responsible for the management of various data-related activities. These data-related activities are performed through different constituents such as data policy and governance, data strategies, data standards, data quality and issues management, data innovation, and analytics engineering (Figure 1.4). The *data policy and governance* constituent is especially important,

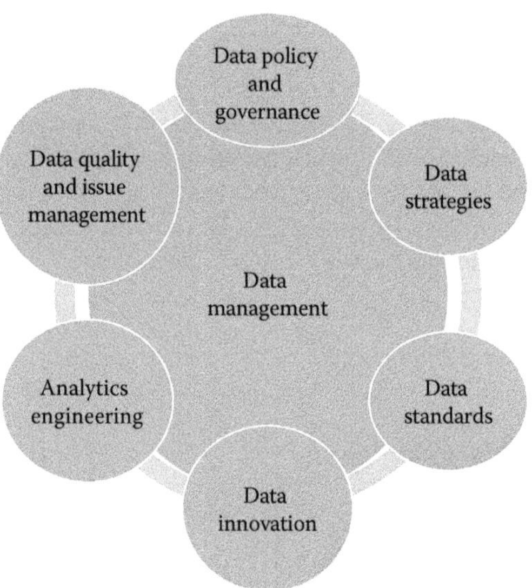

FIGURE 1.4 A typical data management function.

since this will navigate the data-related activities by enforcing data management policies. This item includes steering committees, program management projects and changes in management, data policy compliance, and so on. The *data strategy* constituent is useful in understanding the data and in planning how to use them effectively so that they are fit for the intended purpose. The *data standards* constituent is responsible for ensuring that data have the same meaning and understanding across the organization. The *data quality and issues management* constituent is responsible for cleaning and profiling the data, so that they can be ready for use in various decision-making activities. Ideally, the data quality and the data strategy constituents should work very closely. The *data innovation* constituent is responsible for the systematic use of data analytics to derive meaningful insights and to create value for the enterprise. The last constituent, *analytics engineering*, is responsible for looking at the overall process of executing analytics so that a high quality of analytics is always maintained. The data management function should work closely with various other functions, business units, and technology groups across the organization to create value through data.

An effective data management function should focus on the following important attributes:

- The alignment of data management objectives with the overall organization's objectives in conjunction with strong leadership and support from senior management
- The formation of an effective data quality approach to ensure data are fit for the intended purpose
- The establishment of a sound data quality monitoring and controlling mechanism with an effective issues management system

In Chapter 2, we provide descriptions regarding measuring data quality and the requirements of a data management function.

The next section focuses on the importance of process quality. Process quality is usually associated with the quality of the products that are produced through certain processes. With the introduction of the Six Sigma quality approach and QE approach through Taguchi's methods, several companies have included process quality as an important ingredient in their organizational strategy. In the next section, we also briefly talk about the Six Sigma approach and QE approach. These discussions are sufficient to build the case for the need of a holistic approach for overall robust quality.

1.3 IMPORTANCE OF PROCESS QUALITY

Six Sigma Methodologies

The Six Sigma is a process-oriented approach that helps companies to increase customer satisfaction through drastic improvement of operational performance by minimizing waste and maximizing process efficiency and effectiveness via the use of a set of managerial, engineering, and analytical or statistical concepts. As described in Jugulum and Samuel (2008), Motorola (Chicago, IL, USA) first deployed Six Sigma improvement activities to gain a competitive advantage. Subsequently, with the successful deployment of Six Sigma in companies like General Electric (Boston, MA, USA) and the Bank of America (Charlotte, NC, USA), other organizations started using Six Sigma methodologies extensively. Six Sigma methodologies aim to reduce variations in products/systems/services. The philosophy of Six Sigma is based on the concept of variation. As we know, the quality loss is directly proportional to the variation. Deming (1993) aptly relates the concept of variation with life by stating that variation is life, or life is variation. Therefore, it is very important to understand the sources of variation so that we can act on them and make the products as much alike to each other as possible. Variation can come from several factors such as methods, feeds, humans, and measurements. Based on the sources of variation, there can be two types of variation: (1) variation due to assignable causes or special causes and (2) variation due to chance causes or natural causes.

Special cause variation usually follows a pattern that changes over time (i.e., it is unpredictable). Common cause variation is a stable or consistent pattern of predictable variation over time. The Six Sigma methodologies help us to understand the sources of variation, identify root cause drivers, and optimize processes by reducing the effects of the variation. As mentioned earlier, the Six Sigma approach has been successfully applied in all types of industries including banking, manufacturing, retail, health care, and information technology.

Development of Six Sigma Methodologies

The Six Sigma problem-solving approach is based on five phases: *Define*, *Measure*, *Analyze*, *Improve*, and *Control*, which are collectively known as DMAIC. The Six Sigma concept strives to achieve only 3.4 defects per million opportunities (almost zero defects). Many companies view the Six Sigma approach as a strategic enterprise initiative to improve performance levels by minimizing variation. The concept of 3.4 defects per million opportunities is explained in Figure 1.5.

The calculations in Figure 1.5 are based on normal distribution, as most quality characteristics are assumed to follow the normal distribution. This assumption is valid in the data-driven world, as, in it, we often deal with large samples and,

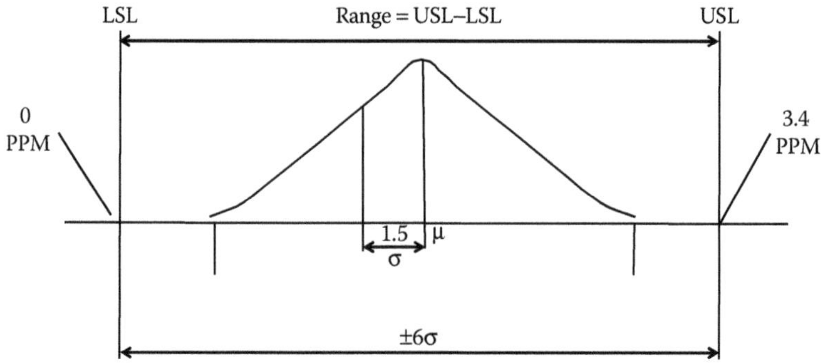

FIGURE 1.5 Concept of variation and sigma level.

for large samples, many distributions can be approximated to normal distribution. In an ideal world, the normal distribution is a bell-shaped curve that is symmetric around the mean. In an ideal normal distribution mean, the median and mode coincide. In addition, in an ideal case, the quality characteristics are perfectly situated at the mean or nominal value (μ). However, in reality, they tend to deviate from this value. There is empirical evidence (McFadden, 1993) that, in a given process, the mean shifts by 1.5 times sigma. For this reason, all defect calculations are based on a 1.5 sigma shift from the mean or nominal value. Table 1.1 shows the impact of defects on sigma levels. This table also emphasizes the importance of achieving Six Sigma quality standards.

Six Sigma projects are executed through the five DMAIC phases. The Six Sigma methodology can be carried out in the steps shown in Figure 1.6.

The DMAIC-based approach is usually applied for existing process improvement activities, for example reducing defects or increasing efficiency. If one needs to design a product or process from scratch, or if an existing process needs a major redesign, it is recommended that a *Design for Six Sigma* (DFSS) approach be used instead. This DFSS methodology is also called the *Define, Measure, Analyze, Design, and Verify* (DMADV) approach. The five phases of DMADV are described in more detail in Figure 1.7.

TABLE 1.1
Magnitude of Sigma Levels[a]

PPM	Sigma Level	PPM	Sigma Level	PPM	Sigma Level
1	6.27	100	5.22	10,000	3.83
2	6.12	200	5.04	20,000	3.55
3.4	**6**	300	4.93	30,000	3.38
4	5.97	400	4.85	40,000	3.25
5	5.91	500	4.79	50,000	3.14
6	5.88	600	4.74	60,000	3.05
7	5.84	700	4.69	70,000	2.98
8	5.82	800	4.66	80,000	2.91
9	5.78	900	4.62	90,000	2.84
10	5.77	1,000	4.59	100,000	2.78
20	5.61	2,000	4.38	200,000	2.34
30	5.51	3,000	4.25	300,000	2.02
40	5.44	4,000	4.15	400,000	1.75
50	5.39	5,000	4.08	500,000	1.5
60	5.35	6,000	4.01		
70	5.31	7,000	3.96		
80	5.27	8,000	3.91		
90	5.25	9,000	3.87		

[a] Assumes a process shift of ±1.5 sigma.

PPM: Parts per million opportunities.

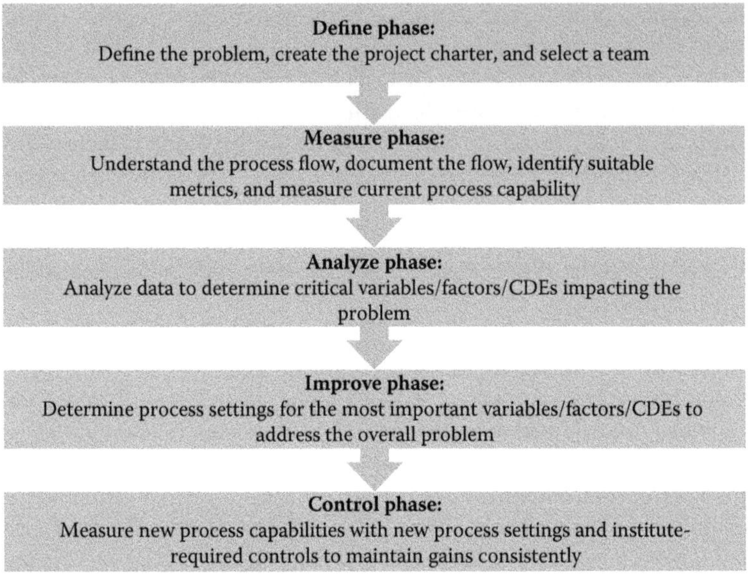

Define phase:
Define the problem, create the project charter, and select a team

Measure phase:
Understand the process flow, document the flow, identify suitable metrics, and measure current process capability

Analyze phase:
Analyze data to determine critical variables/factors/CDEs impacting the problem

Improve phase:
Determine process settings for the most important variables/factors/CDEs to address the overall problem

Control phase:
Measure new process capabilities with new process settings and institute-required controls to maintain gains consistently

FIGURE 1.6 Six Sigma (DMAIC) methodology.

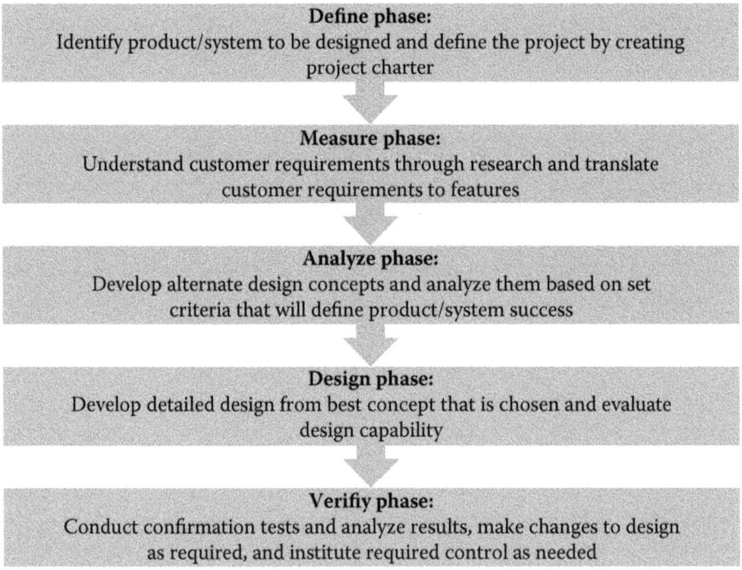

Define phase:
Identify product/system to be designed and define the project by creating project charter

Measure phase:
Understand customer requirements through research and translate customer requirements to features

Analyze phase:
Develop alternate design concepts and analyze them based on set criteria that will define product/system success

Design phase:
Develop detailed design from best concept that is chosen and evaluate design capability

Verifiy phase:
Conduct confirmation tests and analyze results, make changes to design as required, and institute required control as needed

FIGURE 1.7 DFSS (DMADV) methodology.

Process Improvements through Lean Principles

Similar to the Six Sigma process improvement approach, lean principles are also used to improve the processes. However, the main theme behind a lean approach is to improve process speed and reduce costs by eliminating waste.

Womack and Jones (1996) provide a detailed discussion of the lean approach. The five basic principles of this concept are:

1. *Value*: Specify value for the customer
2. *Value stream*: Identify all of the steps in the process and stratify value adds and non-value adds to eliminate waste
3. *Flow*: Allow the value to flow without interruption
4. *Pull*: Let the customer pull value from the process
5. Continuously improve the process for excellence

Process Quality Based on Quality Engineering or Taguchi Approach

Taguchi's quality engineering approach is aimed at designing a product/system/service in such a way that its performance is constant across all customer usage conditions. This approach is considered quite powerful and cost-effective, as it is aimed at improving a product's performance by reducing the variability across various customer usage conditions. These methods have received worldwide recognition both in industry and the academic community, as they have helped to improve companies' competitive positions in the market. In Taguchi's QE approach, there are two types of quality:

1. Customer quality
2. Engineering quality

Customer quality focuses on product features such as color, size, appearance, and function. This aspect of quality is directly proportional to the size of the market segment. As customer quality gets better and better, the companies focusing on such will have a competitive advantage with the possible creation of a new market.

Engineering quality focuses on errors and functional failures. Making improvements to performance or functionality also helps to improve the customer quality. According to Taguchi, this aspect of quality helps in winning the market share because of consistent product performance. Taguchi's QE approach is aimed at improving the engineered quality.

In Taguchi's approach, a signal-to-noise ratio metric is used to determine the magnitude of the product functionality by determining the true output after making the necessary adjustments for uncontrollable or noise variation. Usually, the system is required to perform a set of activities to produce/deliver an intended output by minimizing variations due to noise factors. The output delivery is usually studied by understanding the energy transformation that happens from input to output.

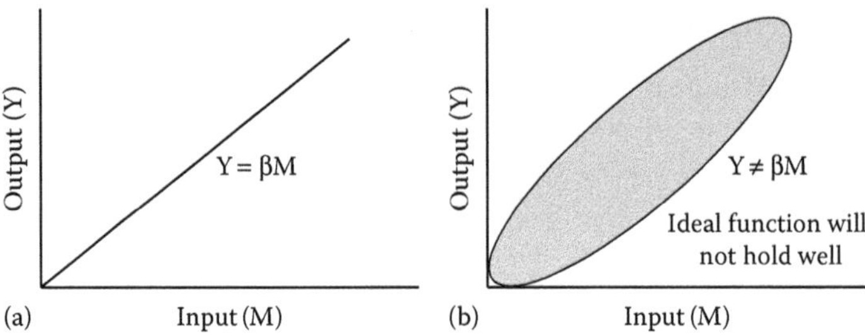

FIGURE 1.8 (a) Ideal function and (b) actual function.

The relationship between the input and the output that governs the energy transformation is often referred to as the ideal functional relationship. When we are attempting to improve product quality, the deviation between the ideal function to actual function is measured through signal-to-noise ratios. This deviation is proportional to the effect of noise factors and, hence, efforts should be made to bring the actual function close to the ideal. This will help in making the performance of products increasingly consistent. This is shown in Figure 1.8. If the rate of energy transformation is perfect or 100% efficient, then there will be no energy losses and so there will be no performance issues or functional failures. However, reality always presents a different picture and, so, efforts should be made to improve energy transformation to increase efficiency levels.

1.4 INTEGRATION OF PROCESS ENGINEERING AND DATA SCIENCE FOR ROBUST QUALITY

From the previous discussions, it is clear that both data quality and process quality aspects are very important. These quality levels should be maintained at high values to improve overall quality. It is not useful to measure metrics related to process quality without knowing if the data used to produce those metrics are of good quality. Since the data field is growing quite rapidly, it is highly important to measure data quality first and then measure process quality to obtain an assessment on overall quality. The main aim of this book is to provide a framework that integrates different aspects of quality by bringing together data science and process engineering disciplines.

Harrington (2006) highlights the importance of managing processes, projects, change, knowledge, and resources for organizational excellence. In addition to these, we need to have the capability to ensure high-quality data and the ability to perform high-quality analytics to derive meaningful outcomes that will help in producing

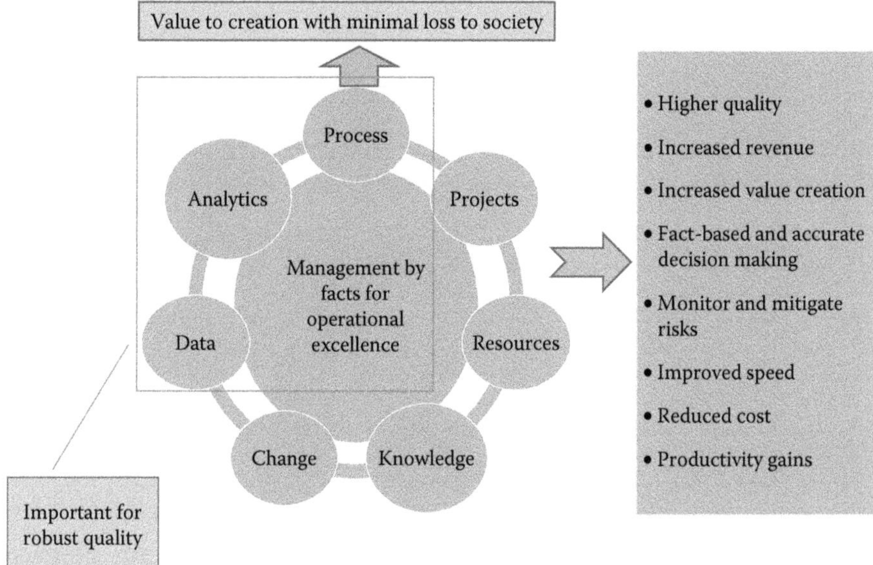

FIGURE 1.9 Important organizational levers for robust quality. (The author thanks John Wiley & Sons for granting permission).

high-quality products. Because of the importance of data and analytics to derive insightful business outcomes, data management and analytics capability management have become critical functions to make sound business decisions and drive business excellence. Figure 1.9 shows the seven levers of a disciplined and effective organization, including the important levers (data, analytics, and process) for robust quality.

2 Data Science and Process Engineering Concepts

2.1 INTRODUCTION

As has been said, the focus of this book is to provide a framework for achieving robust quality by combining different aspects of quality. Many companies have realized the importance of data and are viewing data as key asset in conjunction with other resources including processes. However, some companies are failing to recognize the relationship between data quality (DQ) and process quality (PQ) and often still operate with a silo mentality. There is a need to consider the holistic approach by combining these two powerful aspects of quality. The combined approach uses data science and different process engineering philosophies. This chapter describes some data science and process engineering approaches as they relate to quality concepts in detail for the purpose of facilitating a discussion of integrated approaches.

2.2 THE DATA QUALITY PROGRAM

A good DQ program should satisfy various requirements that will ensure that data are fit for their intended purpose and are of a high quality. This requires a disciplined DQ program that can be applied across the organization. The typical DQ program needs to be focused on building and institutionalizing processes that drive business value and promote a good impact on society.

Data Quality Capabilities

Any DQ program should focus on six important capabilities, as shown in Figure 2.1, among other things. These capabilities will also increase the effectiveness and efficiency of the company's operations.

Strategy and governance: This includes a plan for understanding the current state of the DQ of critical data, the DQ's current level as compared with the target, and how to improve DQ by reducing this gap to meet the strategic goals of the enterprise using a good governance structure.

DQ resources: DQ resources include relevant roles generally filled by skilled people who are capable of executing the DQ program by understating the data and corresponding processes.

Technology infrastructure: This includes the methods, tools, infrastructure, and platforms required to collect, analyze, manage, and report the information related to the organization's critical data.

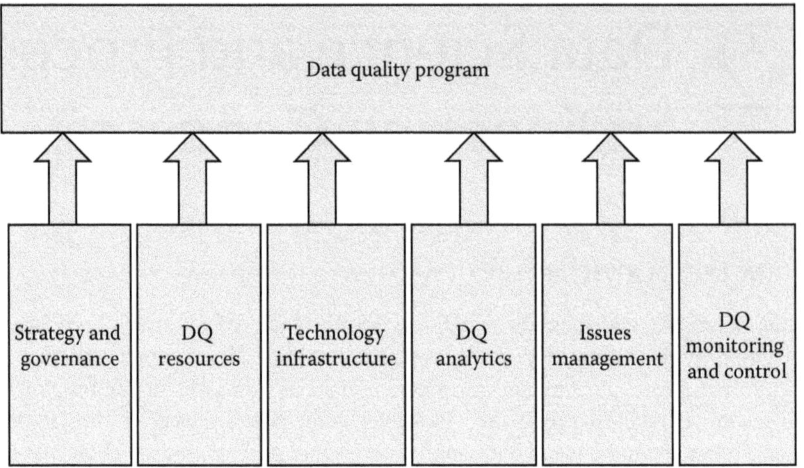

FIGURE 2.1 DQ capabilities.

DQ analytics: This includes the processes, techniques, and tools needed to measure the quality levels of critical data and to conduct root cause analysis for issues to facilitate management processes and problem-solving efforts.

Issues management: The issues management process consists of the identification, tracking, and updating of DQ issues. This includes root cause analysis and remediation efforts that are executed through a governance process. The findings of DQ analytics are usually very important in the management of issues.

DQ monitoring and control: DQ monitoring and control include ongoing activities to measure and control DQ levels and the impact(s) of issues. These activities include establishing a monitoring and control environment; formalizing the change management process; developing scorecards, dashboards, and reports; and putting control processes into place. Typically, statistical process control (SPC) charts are used in monitoring and controlling efforts.

It is important for an organization to implement strong DQ capabilities with a disciplined and standardized approach in order to bring about successful program execution.

2.3 STRUCTURED DATA QUALITY PROBLEM-SOLVING APPROACH[1]

In this section, we will describe a structured DQ approach composed of four phases designed to solve DQ problems or issues. This approach is based on the phases of Define, Assess, Improve, and Control (DAIC) and so is sometimes referred to as the data quality problem-solving approach. This comprehensive approach is aimed at building the best practices and processes for DQ measurement and improvement.

---

1 The author sincerely thanks Brian Bramson for his involvement during this effort.

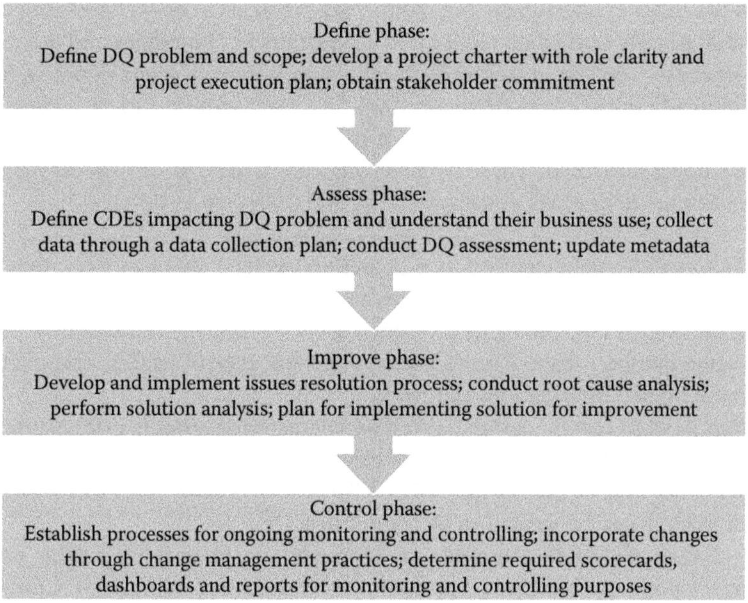

Define phase:
Define DQ problem and scope; develop a project charter with role clarity and project execution plan; obtain stakeholder commitment

↓

Assess phase:
Define CDEs impacting DQ problem and understand their business use; collect data through a data collection plan; conduct DQ assessment; update metadata

↓

Improve phase:
Develop and implement issues resolution process; conduct root cause analysis; perform solution analysis; plan for implementing solution for improvement

↓

Control phase:
Establish processes for ongoing monitoring and controlling; incorporate changes through change management practices; determine required scorecards, dashboards and reports for monitoring and controlling purposes

FIGURE 2.2 DAIC-based DQ methodology. (This was published in Jugulum (2014). The author thanks Wiley for granting permission.)

This strategy is constructed by leveraging Six Sigma approaches [such as Define, Measure, Analyze, Improve, and Control (DMAIC) and Design for Six Sigma (DFSS)/Define, Measure, Analyze, Design, and Verify (DMADV)] to ensure good program or project execution. A brief discussion on Six Sigma approaches was provided in Chapter 1. The DAIC-based DQ methodology and its phases are illustrated with the help of Figure 2.2.

The Define Phase

The Define phase focuses on the definition of the problem(s) by establishing the scope, objectives, resources needed, and project plans with strong governance and stakeholder support. The most important activity in this phase is the creation of the project charter, which will formally establish the scope, objectives, resources, expected business value, and role clarity. The project managers need to prepare a detailed project plan for the four phases of DAIC with all of the relevant tasks and associated deliverables listed.

The Assess Phase

In the Assess phase, the focus will be on defining business use for the critical data and on establishing and assessing a DQ baseline. Further, it will be important to narrow the focus of assessment and monitoring practices to only concerning critical data—that is, only those data required to support the key outcomes and deliverables of the business. The size and complexity of a large company's data population make it economically infeasible to carry out 100% DQ checks for all data elements for any ongoing operational process. Therefore, it is important to reduce the number of data elements being measured. These data elements are termed as critical data elements (CDEs). Formal definitions of data elements and CDEs are as follows:

Data elements: Data elements can be defined as data characteristics that are required in business activities. These elements can take on varying values. Examples of data elements are social security numbers, customer identification numbers, date of birth, account balance, type of facility, commercial bank branch information, and so on.

CDEs: A CDE can be defined as a data characteristic that is *critical* for the organization to be able to conduct business. Typically, a CDE is important for a domain of the business. If a CDE impacts more than one business domain, it can be termed as an enterprise CDE.

Jugulum (2014) provides an approach for identifying CDEs using a scientific prioritization approach. To reduce the number of CDEs, a sampling-based statistical analysis, funnel methodology, is recommended. By applying the funnel methodology, we can reduce the number of data elements by using correlation and regression analyses for continuous data and by using association analysis for discrete data. This application allows us to identify CDEs that have close relationships. Next, a signal-to-noise (S/N) ratio analysis is conducted for each pair of highly correlated CDEs. The CDEs with lower S/N ratios are chosen for future assessment, as a lower S/N ratio indicates a higher variation due to the presence of noise. This means that the process that generates this data is not stable and might impact the smooth functioning of the business and therefore needs attention. Figure 2.3 illustrates this approach.

In Figure 2.3, the approach begins with a set of data elements that we have after collecting input from subject matter experts. An example of funneling CDEs is shown on the right side of this figure. Let us suppose that we have 100 data elements before applying the funnel methodology. After prioritization, through rationalization analysis and a prioritization method like pareto analysis or any other method, we might reduce the number of CDEs to 60. After conducting further statistical analyses, such as a correlation and association analysis, this number can be reduced to 30. Then, by the application of an S/N ratio analysis, this can further be reduced to a manageable list of 15. These 15 data elements are the CDEs that are most important and DQ assessment is performed on them. DQ assessment is generally done by using data profiling techniques and computing descriptive statistics to understand distributional aspects or patterns associated with CDEs. Data profiling includes a basic analysis of missing values, validity against known reference sources, and accuracy as well as data formatting checks.

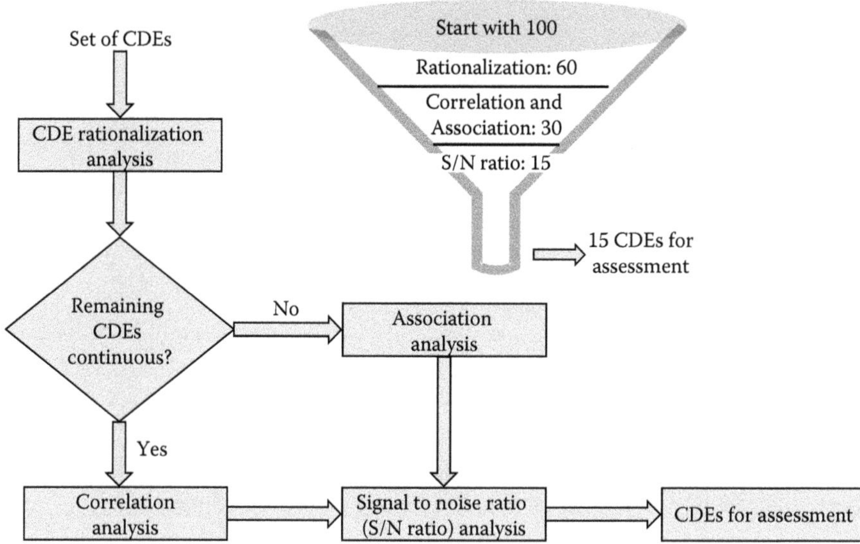

FIGURE 2.3 CDE reduction through the funnel approach (with an illustrative example). (The author would like to thank Chuan Shi for his involvement during this effort.)

DQ dimensions: DQ dimensions are associated with CDEs or data elements and are used to express the quality of data. Wang and Strong (1996) define DQ dimension as a set of attributes that represent a single aspect of DQ. The following four DQ dimensions are commonly used and have been tested by various rules against datasets:

- Completeness
- Conformity
- Validity
- Accuracy

Note that these dimensions are hierarchical; this means that higher-level dimensions such as completeness impact lower-level dimensions. In this regard, accuracy is dependent on validity, validity is dependent on conformity, and conformity is dependent on completeness. It can also be said from this that data, being accurate, do not have meaning if they are not valid; data validity does not have meaning if the data do not conform to a specific format; and conformity is meaningless if the data are not complete.

Table 2.1 shows descriptions of these four DQ dimensions.

Measuring Data Quality

To measure the DQ levels for CDEs, we need to select the DQ dimensions of relevance to the specific business process. Typically, the four DQ dimensions described in Table 2.1 are used. After selecting DQ dimensions, business rules are applied to

TABLE 2.1
Commonly Used DQ Dimensions

Dimension	Definition
Completeness	Completeness is a measure of the presence of core data elements that is required in order to complete a given business process.
Conformity	Conformity is defined as a measure of a data element's adherence to required formats (e.g., data types, field lengths).
Validity	Validity is defined as the extent to which data correspond to valid values as defined in authoritative sources.
Accuracy	Accuracy is defined as a measure of correctness of a data element as viewed in a valid real-world source.

to obtain dimensional-level scores. By applying the DQ rules to CDEs, we classify them as acceptable or unacceptable in the context of a chosen dimension.

Measurement of Data Quality Scores

After selecting the DQ dimensions and measuring them using associated rules, DQ scores are obtained. DQ scores measure the data performance and will indicate if the data are fit for the purpose. A DQ score can be a score for a given DQ dimension; an aggregated score of multiple DQ dimensions of a CDE; or even an aggregated score of multiple CDEs at either the taxonomy level, the function/business unit level, or the enterprise level. A DQ score is expressed as a percentage and so lies between 0 and 100.

DQ scores at multiple levels are usually computed in a sequence; often, they are computed at the dimensional level first, then at the CDE level, and then at the taxonomy level followed by the function/business unit level. After this step, the enterprise-level DQ score can be computed. Figure 2.4 shows this sequence of calculations to be performed.

Figure 2.5 describes the process of obtaining DQ scores at various levels. The approach in Figure 2.5 is based on giving equal weights to all dimensions, CDEs, taxonomies, and functions. If the weights are different, then weighted averages can be used at all levels.

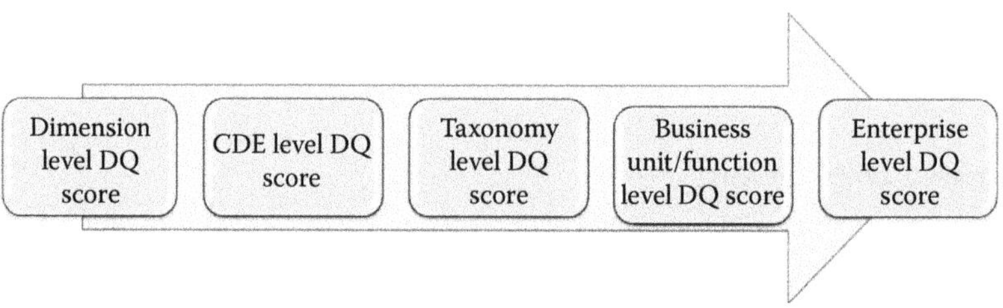

FIGURE 2.4 Measuring enterprise-level DQ score.

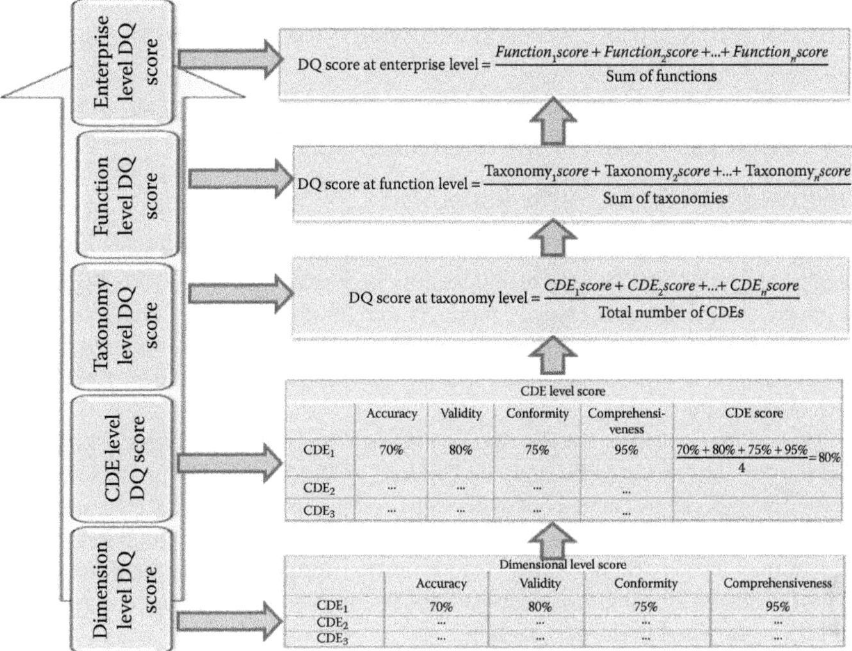

FIGURE 2.5 DQ scores at various levels. (This was published in Jugulum (2014). The author thanks Wiley for granting permission.)

The Improve Phase

The Improve phase of the DAIC approach focuses on improvement activities based on the DQ results from the Assess phase. The issues causing lower DQ scores must be identified and root cause analysis must be carried out. The issues management system should contain these issues with their severity levels (usually with respect to a scale of low, medium, or high priority). The data, technology, and business teams should work together to conduct root cause analysis on issues and resolve them. There should be a robust issue management system to manage issues and resolve them. Remediation and improvement efforts have to be institutionalized and constantly monitored to ensure that management teams and executives have visibility of the DQ improvement.

The Control Phase

This last phase of the DAIC approach is aimed at monitoring and controlling the improvement activities with scorecards, dashboards, control, charts, and so on. During this phase, scorecards and dashboards are created as part of ongoing control and monitoring processes. The ongoing monitoring and improvement activities should include statistical process control aspects for monitoring.

The main objective of the DAIC approach is to ensure data are fit for the intended purpose by gaining control over the key data used in business processes so that effective decisions can be made.

2.4 PROCESS QUALITY METHODOLOGIES

Development of Six Sigma Methodologies

Six Sigma was first developed as a statistics-based methodology to *Define*, *Measure*, *Analyze*, *Improve*, and *Control* manufacturing processes. The goal of Six Sigma was to improve or design processes or products to reduce defects (Six Sigma being the measure of 3.4 defects per million opportunities). Over a period of time, Six Sigma has evolved as a strategic approach that organizations can use to gain competitive advantages in the market.

As mentioned before, the Six Sigma approach is a process-driven methodology. The projects are executed through the DMAIC or DFSS/DMADV processes. These methodologies can be carried out in the steps as described in Chapter 1. Jugulum and Samuel (2008) have additionally described a Six Sigma-related methodology called Design for Lean Six Sigma (DFLSS) to improve overall quality in the development of new products by combining lean and Six Sigma principles. A brief description of DFLSS is provided in the next section.

Design for Lean Six Sigma Methodology

When the DFSS approach is combined with lean principles, this methodology is referred to as DFLSS. Figure 2.6 describes the DFLSS methodology. In DFLSS,

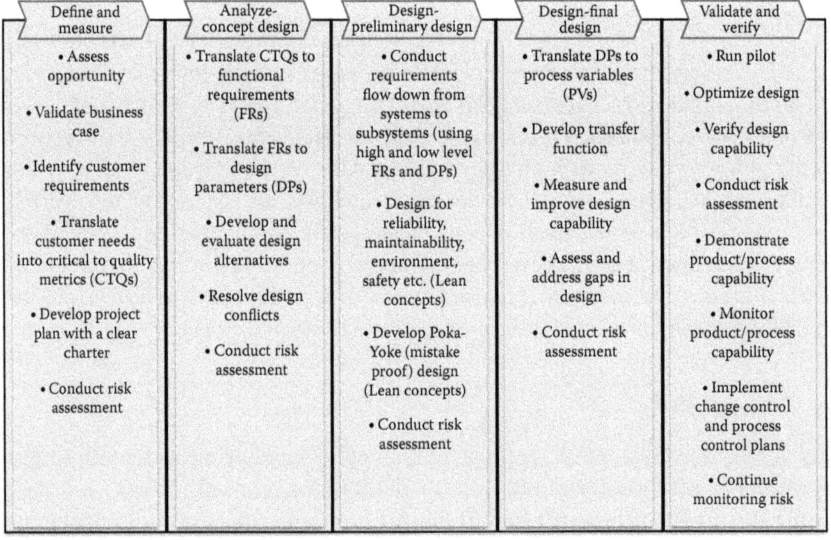

FIGURE 2.6 DFLSS methodology.

various engineering, quality, and statistical concepts, tools, and techniques are integrated and used in a systematic fashion to achieve Six Sigma quality levels.

As can be seen from Figure 2.6, the DFLSS road map is built in accordance with the DMADV methodology. This approach which covers all of the requirements of DMADV and is aligned with the main steps of DMADV is as follows:

- *Define and measure*: *Identify customer needs by assessing opportunity*. In this step, a feasibility assessment needs to be conducted to validate the business case by clearly defining objectives.
- *Analyze*: *Deliver the detailed concept design by evaluating various design alternatives and addressing design conflicts*. The concept design stage is one of the most important stages of the DFLSS method. This is the phase wherein the conversion of customer needs to actionable and measurable metrics (also referred to as critical to quality metrics or CTQs) takes place. In this phase, the requirements are transferred to lower levels so that the design requirements can be understood better and good concepts can be developed. Dr. Genichi Taguchi's concept design approach or other strategies can also be used. We may also consider the Pugh concept selection approach to select the best alternative against required criteria such as cost, simplicity, or cycle time.
- *Design*: *Develop an initial design and final design by including lean concepts*. In the preliminary design stage, a flow-down approach and robust design strategies may be used. The final design stage is a very important step in which the design from a productivity, reliability, and quality point of view is developed. The development of transfer function, the use of Taguchi methods, and two-step optimization are all very important for the purpose of optimization and for getting to the final design.
- *Validate and verify*: *Validate the design and verify the capability*. This is the phase during which the final design is tested against performance and capability predictions. A pilot design is created and a confirmation run is conducted to validate the performance of the design. In addition, the process used to build the product is also validated. The final design is brought to actual practice and the results of the design activities are implemented. The design, measurement controls, and process controls are institutionalized with suitable control plans. Finally, the results of the product design are leveraged with other applications. It is important to note that, at all stages, the risk is constantly evaluated so that required actions can be taken as necessary.

2.5 TAGUCHI'S QUALITY ENGINEERING APPROACH

Taguchi's quality engineering approach is aimed at designing a product or service in such a way that its performance is the same across all customer usage conditions. Taguchi's methods of quality are aimed at improving the functionality of a product or service by reducing variability across the domain of customer usage conditions. They are considered to be powerful and cost-effective methods.

These methods have received recognition across the globe both in industry and the academic community.

Engineering Quality

Taguchi's approach[2] of quality engineering is based on two aspects of quality:

1. Customer quality
2. Engineering quality

The second aspect, engineering quality, addresses defects, failures, noise, vibrations, and pollution, among others. Taguchi methods aim to improve this aspect of quality.

Engineered quality is usually influenced by the presence of three types of uncontrollable or noise factors; specifically, these are:

1. Customer usage and environmental conditions
2. Product wear and deterioration
3. Manufacturing imperfections and differences that occur among individual products during manufacturing

A typical product development initiative is performed in three stages though most applications using Taguchi methods focused on parameter design. These stages are:

a. Concept design
b. Parameter design
c. Tolerance design

Methods based on Taguchi's approach are developed with the following principles:

1. Evaluation of the functional quality through energy transformation
2. Comprehension of the interactions between control and noise factors
3. Use of orthogonal arrays (OAs) for conducting experiments
4. Use of signal-to-noise ratios to measure performance
5. Execution of two-step optimization
6. Establishment of a tolerance design for setting up tolerances

We will briefly describe these principles in the following. For more information, please refer to Taguchi (1987) and Phadke (1989).

Evaluation of Functional Quality through Energy Transformation

To begin with, Taguchi methods focus on identifying a suitable function (called an ideal functional relationship) that governs the performance of the system. The ideal function

[2] The author is grateful to Dr. Genichi Taguchi for allowing the use of his materials on quality engineering.

helps in understanding the energy transformation in the system by evaluating useful energy (i.e., the energy that is successfully used to get the desired output) and wasteful energy (i.e., the energy spent because of the presence of uncontrollable or noise factors). The energy transformation is measured in terms of S/N ratios. A higher S/N ratio means lower effect of noise factors and also implies efficient energy transformation.

Understanding the Interactions between Control and Noise Factors

In the Taguchi quality engineering approach, the control factors are adjusted to minimize the impact of noise factors on the output. Therefore, it is important to understand the interactions between control and noise factors. In other words, the impact of the combined effect of control and noise factors must be studied to improve performance.

Use of Orthogonal Arrays

OAs are used to study various combinations of factors in the presence of noise factors. OAs help to minimize the number of runs (or combinations) needed for the experiment by taking only a fraction of the overall experimental combinations. These combinations are based on specific conditions that are required to be studied to understand the effects of factors on the output. For each combination of OAs, the experimental outputs are generated and analyzed.

Use of Signal-to-Noise Ratios to Measure Performance

A very important contribution of Dr. Taguchi to the quality world is the development of S/N ratios for measuring system performance. The S/N ratio is used to determine the magnitude of true output (transmitted from the input signals) in the presence of uncontrollable variation (due to noise). In other words, the S/N ratio measures the effectiveness of energy transformation. The data generated through OA experiments are analyzed by computing S/N ratios. The S/N ratio analysis is used to make decisions about optimal parameter settings.

Two-Step Optimization

After conducting the experiment, the factor-level combination for the optimal design is selected with the help of two-step optimization. The first step is to minimize the variability (i.e., maximize the S/N ratios). In the second step, the sensitivity (mean) is adjusted to the desired level. According to Dr. Taguchi, it is easier to adjust the mean to the desired level by changing the settings of one or two factors. Therefore, the factors affecting variability must be studied first, as many factors have an effect on variability.

Tolerance Design for Setting up Tolerances

After identifying the best parameter settings using parameter design, tolerancing is done to determine allowable ranges or thresholds for each parameter in the optimal design. The quality loss function approach, which was described in Chapter 1, is usually employed to determine tolerances or thresholds.

Additional Topics in Taguchi's Approach

Parameter Diagram

In the Taguchi approach, a parameter diagram or p-diagram is used to represent a product or a system. The p-diagram captures all of the elements of a process just as a cause and effect diagram or a Suppliers, Inputs, Processes, Outputs, and Customers (SIPOC) diagram does. Figure 2.7 shows all of the required elements of the p-diagram. The energy transformation takes place between the input signal (M) and the output response (y). The goal is to maximize energy transformation by adjusting control factor (C) settings by minimizing the effect of noise factors (N). As mentioned earlier, an S/N ratio is used to measure the effectiveness of energy transformation.

1. *Signal factors* (*M*): These are factors that are selected based on customer usage conditions. Quality improvement efforts are performed based on these factors and they should have a high degree of correlation with the output if the energy transformation is effective. For example, the application of force on a brake pedal is a signal factor for the braking unit of an automobile. Signal factors are typically selected by the engineers based on engineering knowledge and the range of usage conditions to be considered in the design of the product/system.
2. *Control factors* (*C*): As the name suggests, these factors are in the control of the designer or engineer and can be changed. In a p-diagram, only *control factor* elements can be changed by the designer/engineer. The different values that control factors can take are referred to as levels, and appropriate levels should be chosen to improve performance quality.
3. *Noise factors* (*N*): Noise factors are also called uncontrollable factors. They cannot be controlled and their presence in the system affects (in a negative way) the energy transformation from the input to the output. Since these factors cannot be controlled, it is important to adjust the levels of control factors in such a way that product performance is insensitive to noise factors. As mentioned earlier, noise factors can come from customer usage and environmental conditions, product wear and deterioration, and manufacturing imperfections and differences in individual products that occur during manufacturing.

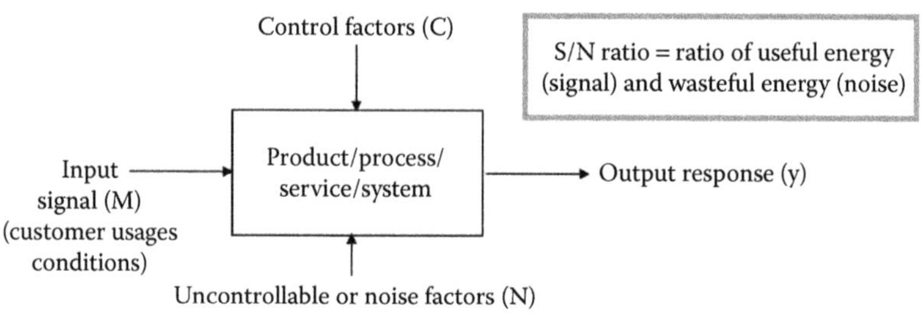

FIGURE 2.7 Elements of a parameter diagram or p-diagram.

4. *Output response* (y): Output response corresponds to the output of energy transformation in the product/system. For example, stopping distance, or the time in which an automobile comes to halt when force is applied on the brake pedal, is the output response of the brake unit of an automobile.

Design of Experiments

The design of experiments (DOE) is a subject that will help in ensuring experiments are conducted in a systematic fashion and in analyzing the results of experiments to find optimal parameter combinations to improve overall product or system performance. The design of experiments is extensively used in many disciplines for optimizing performance levels. There is extensive literature on this subject. A typical experimental design cycle can be seen in Figure 2.8 and consists of the following five important steps:

1. Planning and designing of the experiment
    - Understand the main function(s) to be improved
    - Identify input signal, control factors, noise factors, and output response
    - Design a suitable experiment (full factorial or fractional factorial)
2. Executing the experiments
    - Perform all experimental runs (hardware or simulation-based)
    - Measure output associated with all runs

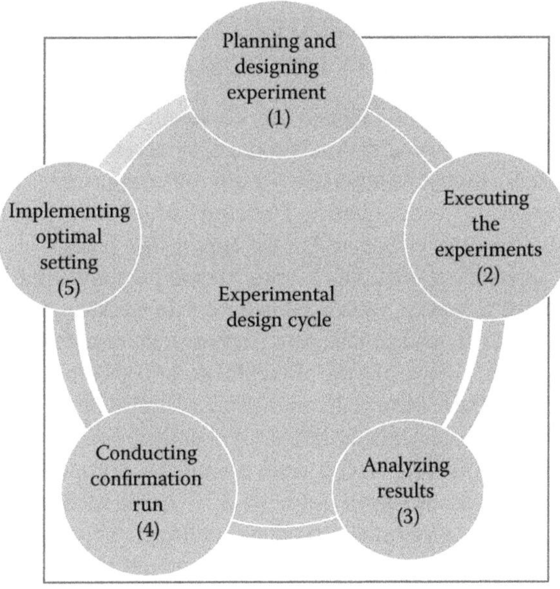

FIGURE 2.8 Experimental design cycle.

3. Analyzing the experimental results
   - Analyze experimental results using suitable techniques
   - Determine optimal setting(s)
   - Predict performance level(s) for optimal setting(s)
4. Conducting confirmation run
   - Conduct a confirmation run as necessary
   - Compare actual performance with predicted performance
5. Implementing optimal settings
   - Implement optimal settings
   - Monitor and control ongoing performance

Types of Experiments

There are typically two classes of experiments:

Full factorial experiments: In a full factorial experiment, all possible combinations of factors are studied and data are collected and analyzed to understand the effect of all factors and all possible interaction or combination effects.

Fractional factorial experiments: If the number of factors to be studied is too large, fractional factorial experiments are used instead, as full factorial experiments might require a significant amount of time, resources, and money. In fractional factorial experiments, a fraction of the total number of experiments is studied. Main effects and important interaction or combination effects are estimated through the analysis of experimental data. The orthogonal arrays belong to this class of experiments.

2.6 IMPORTANCE OF INTEGRATING DATA QUALITY AND PROCESS QUALITY FOR ROBUST QUALITY

It should be now clear that both DQ and process quality aspects are very important in the design and development of new products or services. Whether we use the Six Sigma approach, Taguchi's approach, or any other approach, it is important to include DQ for ensuring overall quality. Therefore, when Six Sigma approaches are used, DQ aspects should be considered in the appropriate phases.

If a DMAIC approach is used, DQ aspects should be considered in the Measure, Improve, and Control phases, as shown in Figure 2.9. In the Measure phase, the DQ of metrics should be ensured, while, in the Analyze phase, the quality of factors/variables/CDEs should be confirmed by using the principles described before. The Control phase should focus on deploying control plans on metrics as well as on variables to make sure that the DQ levels and process quality levels are maintained on an ongoing basis.

If a DFSS approach is used, DQ aspects should be considered in the Measure, Design, and Verify phases, as shown in Figure 2.10. In the Measure phase, the accuracy and reliability of requirements should be confirmed as part of DQ assessment, and, in the Design phase, DQ should be ensured from all sources. The Verify phase should focus on deploying control plans on metrics as well as on variables to make sure that both DQ as well as process quality levels are maintained on an ongoing basis.

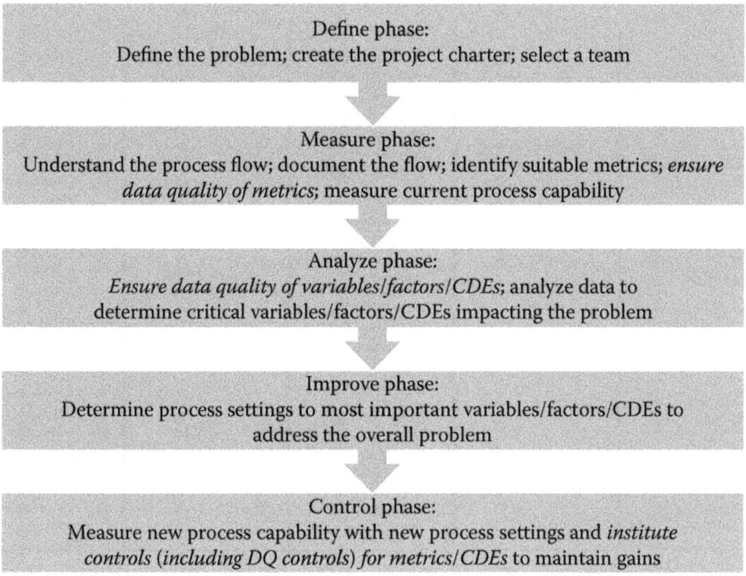

Define phase:
Define the problem; create the project charter; select a team

↓

Measure phase:
Understand the process flow; document the flow; identify suitable metrics; *ensure data quality of metrics*; measure current process capability

↓

Analyze phase:
Ensure data quality of variables/factors/CDEs; analyze data to determine critical variables/factors/CDEs impacting the problem

↓

Improve phase:
Determine process settings to most important variables/factors/CDEs to address the overall problem

↓

Control phase:
Measure new process capability with new process settings and *institute controls (including DQ controls) for metrics/CDEs* to maintain gains

FIGURE 2.9 DMAIC approach for robust quality.

Define phase:
Identify product/system to be designed; define the project by creating project charter

↓

Measure phase:
Understand customer requirements through research; *ensure accuracy and reliability of requirements (DQ aspects)*; translate customer requirements to features

↓

Analyze phase:
Develop alternate design concepts and analyze them based on set criteria that will define product/system success

↓

Design phase:
Ensure DQ from all sources; develop detailed design from best concept that is chosen; evaluate design capability

↓

Verify phase:
Conduct confirmation tests and analyze results; make changes to design as required; *institute required controls (including DQ controls) as needed*

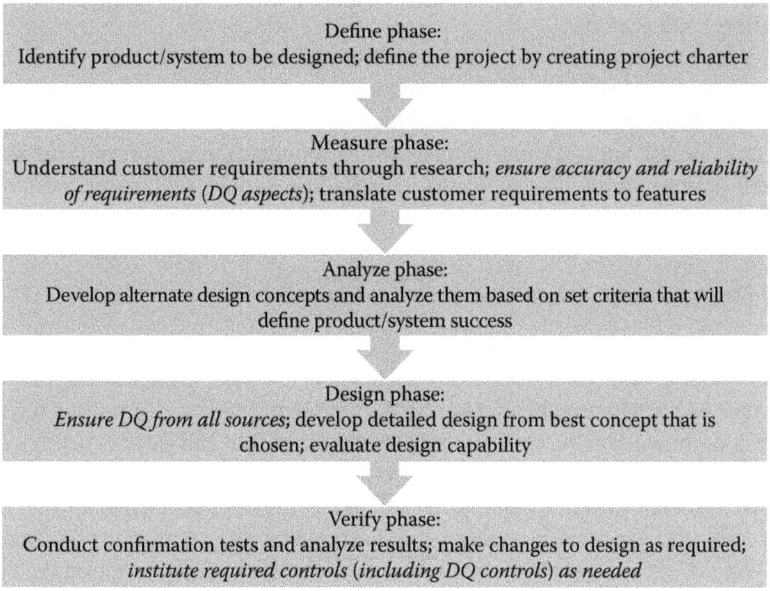

FIGURE 2.10 DFSS approach for robust quality.

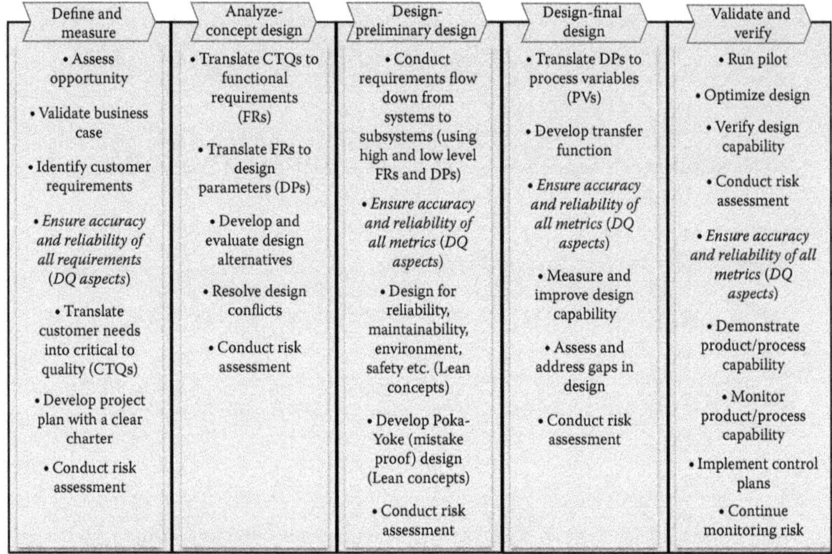

FIGURE 2.11 DFLSS approach for robust quality.

The DFLSS approach (Figure 2.6) should also be modified, as shown in Figure 2.11, to cover the DQ aspects for achieving robust quality.

In the case of Taguchi's methods, the DQ aspects are very important in the parameter design stage wherein the design of experiments is employed. Prior to collecting data, we need to make sure that the data sources, measurement systems, and so on, are reliable and accurate.

As you can see, monitoring and controlling quality levels are important in maintaining robust quality levels. SPC techniques are very useful in this regard. SPC has a distinct advantage over other forms of monitoring, as it is based on numerical facts (i.e., data). A SPC approach can also be used to automate the identification of anomalies and in determining thresholds for metrics and variables/CDEs in combination with subject matter expertise.

Brief Discussion on Statistical Process Control

SPC is a method for measuring the consistency and ensuring the predictability of processes using numerical facts or data. This concept was introduced and pioneered by Walter A. Shewhart in the first half of the twentieth century. Controlling a process makes it more predictable by reducing process variability and clearly distinguishing the causes of variation (both common causes and special causes). The aim of SPC is to understand the variation associated with processes and data elements.

Usually, the variation is measured against customer expectations or specifications. Any deviation from these is undesirable; this thus makes variation the enemy of quality. Therefore, it is very important to understand the sources of variation so that we can act on them and make observations and measurements as consistent as

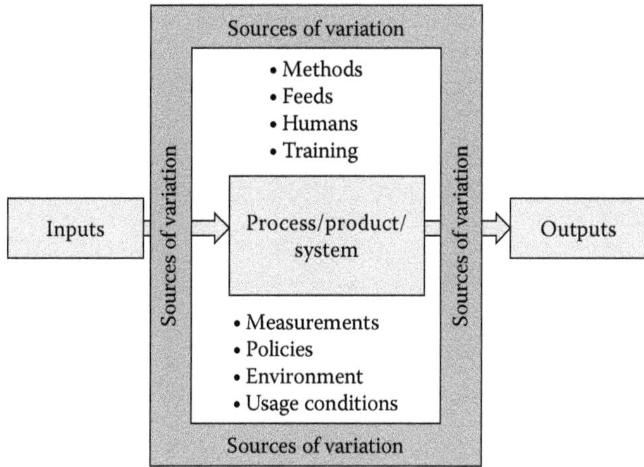

FIGURE 2.12 Sources of variation.

possible. The sources of variation can come from a range of factors such as methods, feeds, humans, and measurements, as shown in Figure 2.12. Figure 2.12 is a typical process representation with inputs, outputs, and sources of variation. Within a given process/system or between processes/systems, these factors will have different effects. Successful SPC deployment requires a detailed understanding of the processes and associated factors that cause variation.

A primary tool for SPC is the control chart: a time series plot or run chart that represents the set of measurements, along with their historical mean and upper and lower control limits (UCLs and LCLs). These limits are three standard deviations (sigma) from the mean of the measurements. SPC control charts help in detecting points of unusual or unexpected variation (i.e., measurements above the UCL or below the LCL). Figure 2.13 shows control charts, along with various components.

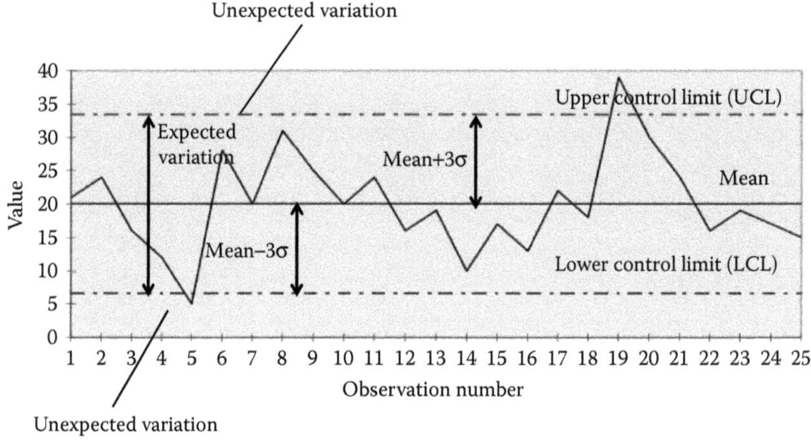

FIGURE 2.13 Control chart and its components.

The control charts illustrate the stability, predictability, and capability of the process via a visual display of variation.

In the next chapter, we discuss the alignment of data and process strategies with the corporate strategy while prioritizing requirements and metrics management. This will be quite helpful in focusing on areas of immediate interest wherein data science and process engineering aspects are absolutely essential to manage metrics and variables.

3 Building Data and Process Strategy and Metrics Management

3.1 INTRODUCTION

As mentioned earlier, organizations have begun to view data as critical assets, giving equal importance to them as to other items such as people, capital, raw materials, and infrastructure. The concept of data as asset has driven the need for dedicated data management programs that are similar in nature to Six Sigma process engineering programs. Beyond ensuring data and processes are fit for the intended business purposes, organizations should focus on the creation of shareholder value through data- and process-related activities. To achieve this, organizations must strive to develop a data and process strategy that includes components such as data monetization, data innovation, risk management, process excellence, process control, and process innovation. Key characteristics of such a strategy should include speed, accuracy, and precision of managing the data and processes to help differentiate the organization from global competitors. This strategy should also be tightly aligned with the corporate strategy so that requirements can be prioritized and executed in a systematic fashion.

This chapter is aimed at discussing the successful design and development of a data and process strategy to create value with simultaneous maintenance of the strategy's alignment with corporate objectives. This chapter also describes how we can decompose the strategy into lower-level components with suitable design parameters (DPs) to address complexity and sequences of execution for resource planning. In addition, the chapter also highlights metrics management aspects that are essential for executing data and process strategies.

3.2 DESIGN AND DEVELOPMENT OF DATA AND PROCESS STRATEGIES

In any organization, data and process strategies play an important role, as they help in understanding the impact of data and processes across the organization and in planning and governing these assets. Generally, data and process strategy requirements should include the following:

Data valuation: Data valuation focuses on the idea of estimating the dollar value for data assets. Many companies are interested in monetizing data. So, data valuation should be an important requirement for overall data and process strategy.

Innovation: Data innovation deals with the systematic use of data and process efficiency techniques to quickly derive meaningful insights and value for the company. Data innovation also helps to provide intelligence about customers, suppliers, and the network of relationships.

Risk management and compliance: Risk management and compliance deals with risk aspects by quantifying the risk of exposure for all legal, regulatory, usage, and privacy requirements for data and processes at various levels.

Data access control: This is needed to ensure that data access, authentication, and authorization requirements for data are met at all levels.

Data exchange: The concept of data exchange helps in understanding internal and external data using standard data definitions and in ensuring the data are fit for the intended purpose.

Monitoring, controlling, and reporting: This is an important requirement for overall strategy, as it helps to provide a real-time reporting mechanism with monitoring and controlling aspects to understand the end-to-end process- and data-related activities by performing real-time analytics to support business decisions.

Build-in quality: This emphasizes the need to institutionalize quality practices and embed them into processes for business self-sufficiency to achieve standardization across an enterprise.

Data as service: This requirement helps in providing seamless, business-friendly access to data services and inventory through enabling technologies.

These strategic requirements are intended to provide the following benefits to the organization:

- New revenue streams
- Increased shareholder value
- Objective, fact-based decision support
- Monitoring and mitigation of data risks
- Improved delivery speed
- Reduced cost
- Increased capacity
- Improved data and insights quality

3.3 ALIGNMENT WITH CORPORATE STRATEGY AND PRIORITIZING THE REQUIREMENTS[1]

After developing data and process strategy requirements, we need to ensure that they are well-aligned also with the corporate strategy requirements, as it is important to ensure that the data and process aspects will help in delivering value to the organization. The use of a methodology, such as that described in Figure 3.1, is very helpful for this alignment.

[1] The author would like to thank Chris Heien for his involvement and help in this effort.

Data and process strategy
Key levers to delivery value
DPS1
DPS2
DPS3
DPS4
DPS5
...

Align corporate priorities with data and process strategy key levers

Data and process strategy	Corporate strategy					
Functional requirement	Capabilities and priority					
	CS1	CS2	CS3	CS4		
	9	5	3	1	Total	Rank
DPS1	9	9	3	1	136	2
DPS2	9	3	3	1	106	5
DPS3	9	9	9	3	156	1
DPS4	3	9	9	9	108	4
DPS5	9	9	1	3	132	3
...	...	...	...	...	...	...

FIGURE 3.1 Alignment of data and process strategy with the corporate strategy (illustrative only).

Figure 3.1 describes the process of alignment with the help of a prioritization matrix. This matrix is a very helpful tool for prioritizing the proposed data and process strategy requirements based on corporate strategy criteria. At first, we need to list all data and process strategy requirements (DPS1, DPS2…) and corporate strategy requirements. The corporate strategy requirements/enablers (CS1, CS2…) should be ranked based on business criteria according to a scale of 1 to 9 (1 being least important and 9 being most important). After this step, each proposed data and process strategy requirement is scored in relation to each corporate strategy requirement/enabler on a scale of 1, 3, and 9 (1 meaning weak relationship, 3 meaning moderate relationship, and 9 meaning strong relationship). Once we determine both the weights of the corporate strategy requirements/enablers and the scores of the data strategy requirements in relation to those weights, we can compute a total score, which is a sum of the products of all criteria weights and their corresponding scores. The total score, called the alignment index, is used to rank the importance of data and process strategy requirements. In Figure 3.1, as an example, DPS3 is ranked first, DPS1 is ranked second, and so on.

Using this approach, the requirements of data and process strategy listed in the beginning of this section have been ranked against four enablers of corporate strategy. The details of this alignment have been provided in Figure 3.2. The top five data strategy requirements are also highlighted.

Data and process strategy		Corporate strategy					
Key levers		Enabler 1 (9)	Enabler 2 (5)	Enabler 3 (3)	Enabler 4 (1)	Total	Rank
DPS1	Data valuation	3	9	3	3	84	7
DPS2	Data innovation	9	9	3	1	136	2
DPS3	Decision support	9	9	3	1	136	3
DPS4	Risk management and compliance	9	9	9	9	162	1
DPS5	Data access control	3	9	9	9	108	5
DPS6	Data exchange	3	9	1	3	78	8
DPS7	Monitoring, controlling, and reporting	3	3	3	1	52	9
DPS8	Build-in quality	9	9	1	3	132	4
DPS9	Data as service	9	3	1	3	102	6

FIGURE 3.2 Data and process strategy—corporate strategy alignment matrix.

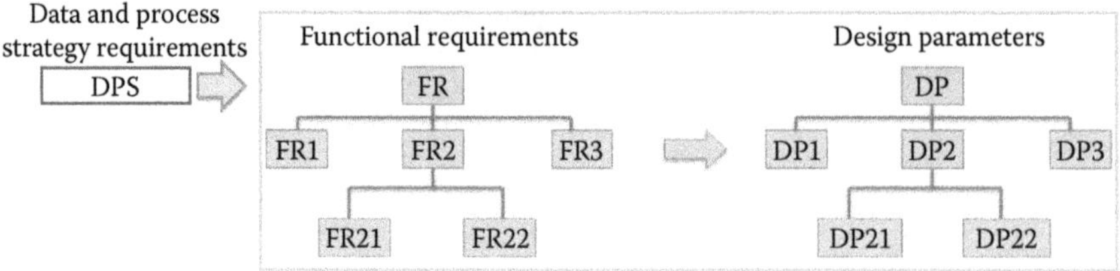

FIGURE 3.3 Identify *how-to-aspects* for requirements.

In this section, the prioritization process of data and process strategy requirements has been discussed. After this, we need to identify *how-to aspects* for these requirements and prioritize them, as shown in Figure 3.3. The prioritized data and process requirements [also referred to as functional requirements (FRs)] can be decomposed into lower-level requirements along with corresponding *how-to aspects* or design parameters (DPs) using the theory of axiomatic design.

In the next section, we discuss the concepts associated with axiomatic design theory. In axiomatic design theory, the FRs constitute the *what* part (what requirements are needed) and the DPs constitute the *how* part (how requirements are satisfied) of the design.

3.4 AXIOMATIC DESIGN APPROACH

The Axiomatic design theory developed by Professor Nam P. Suh has been used in developing systems, processes, or products that related with software, hardware, and manufacturing. This subject (Suh, 2001) has been extensively used for the following purposes:

1. To provide a systematic way of designing products and systems through a clear understanding of customer requirements
2. To determine the best designs from various design alternatives
3. To create a robust system architecture by satisfying all of the FRs

Axiomatic design theory can be used to translate customer requirements to FRs and translate FRs into DPs. Axiomatic design theory is more effective than other techniques since, in axiomatic design, we try to uncouple or decouple the design to satisfy FRs independently of one another. Usually the design is converted into uncoupled or decoupled design by using a systematic flow-down approach with design equations.

Design Axioms

The axiomatic design approach is based on the following two axioms:

1. The independence axiom
2. The information axiom

The independence axiom states that the FRs must always be satisfied independently of each other. The way to achieve this independence is through uncoupling or

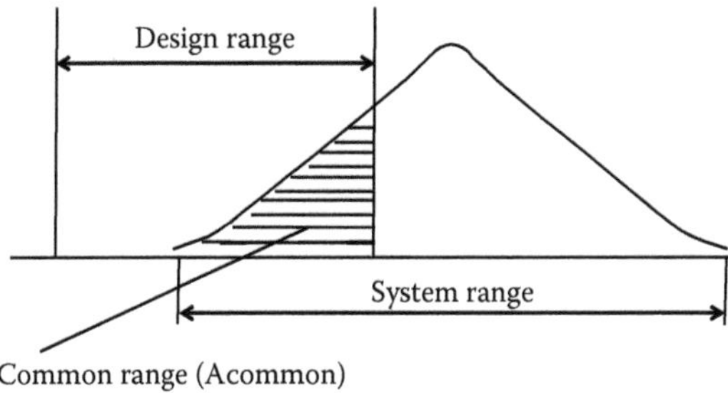

FIGURE 3.4 Explanation of the information axiom.

decoupling the design. FRs are defined as the minimum set of independent requirements that characterize the design objectives (Suh, 2001). The second axiom, the information axiom, states that the best design has the least information content. The information content is expressed in bits and calculated using probability theory. Figure 3.4 explains the concept of the information axiom.

From Figure 3.4, we can say that the probability of success corresponding to an FR is calculated by using a design range or tolerance and system range or variation, or by process variation, as shown in Figure 3.4. The information content is calculated by area under the common range (Acommon) and is expressed in bits. The equation for calculating information content is as given below:

$$\text{Information content} = I = \log_2\left(\frac{1}{\text{Acommon}}\right)$$

From this equation, it is clear that, if Acommon = 1 or the design range is equal to the system range, then the information content is zero, indicating that the design is the best. So, according to axiomatic design theory, any design is good as long as the system range is within the design range. Axiomatic design theory coupled with statistical process control and other variation analysis techniques helps in the selection of the best design with lower variability.

Designing through Domain Interplay

The axiomatic design is based on domain thinking. The involved domains are the customer domain, the functional domain, and the physical domain. Suh (2001) defines the design as interplay between these domains by addressing the design considerations of *what we want to achieve* and *how we achieve it/them*. The domain thinking concept is the basis of axiomatic design theory. A typical domain structure is shown in Figure 3.5. The domain on the left relative to a particular domain represents the concept of *what we want to achieve*, whereas the domain on the right represents the design solution concept of *how we achieve it/them*.

In the customer domain, we capture the needs (or requirements) that the customer is looking for in a product, process, or system. In the functional domain, the customer needs are translated into FRs. The specific FRs are satisfied by identifying

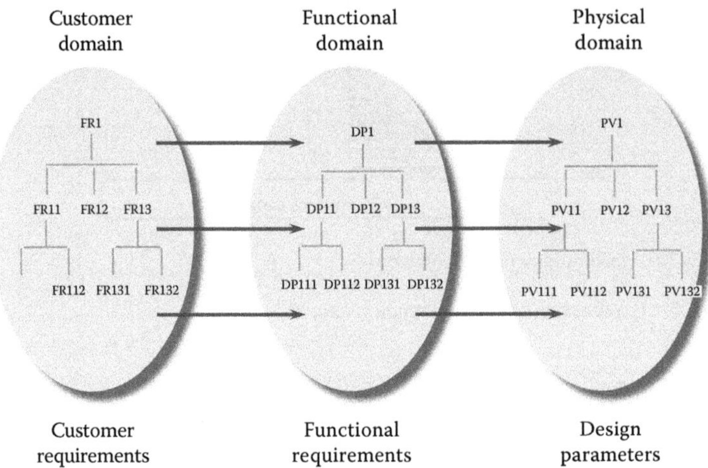

FIGURE 3.5 Designing through domain interplay.

suitable DPs in the physical domain. Typically, the design equations are written in the following matrix representation:

$$\{FR\} = [A]\{DP\}$$

In the above representation, [A] is the design matrix. The elements of design matrix [A] represent sensitivities (i.e., changes in FRs with respect to changes in DPs) and are expressed using partial derivatives.

If we have three FRs and three DPs, the design matrix will be as follows:

$$[A] = \begin{bmatrix} A11 & A12 & A13 \\ A21 & A22 & A23 \\ A31 & A32 & A33 \end{bmatrix}$$

In the case of three FRs and three DPs, we can also have the following linear equations:

$$FR1 = A11\ DP1 + A12\ DP2 + A13\ DP3$$
$$FR2 = A21\ DP1 + A22\ DP2 + A23\ DP3$$
$$FR3 = A31\ DP1 + A32\ DP2 + A33\ DP3$$

Based on the structure of the design matrix, we usually will have three types of designs: uncoupled, decoupled, and coupled. Uncoupled designs are desirable because their design matrix will be a diagonal matrix indicating that every FR can be satisfied by one particular DP, and thus they satisfy the requirements of the independent axiom. If it is not possible to obtain an uncoupled design, then we should try to obtain a decoupled design. The decoupled design matrices have structures of upper or lower triangular matrices. Decoupled designs help us to follow a sequence by which we can fix DPs in a particular order to satisfy the FRs. Satisfying FRs in a particular order helps to satisfy the requirements independently and therefore we can maintain the independence axiom requirements. All other structures of design matrices indicate coupled designs. In a three FR and three DP case, the uncoupled and decoupled design matrices will have the following structure:

$$[A]=\begin{bmatrix} A11 & 0 & 0 \\ 0 & A22 & 0 \\ 0 & 0 & A33 \end{bmatrix} \qquad [A]=\begin{bmatrix} A11 & 0 & 0 \\ A21 & A22 & 0 \\ A31 & A32 & A33 \end{bmatrix}$$

Uncoupled Design Decoupled Design

The concept of axiomatic design helps to us move from one domain to another (e.g., from the customer domain to the physical domain) so that we can decompose the requirements to the lower levels in a systematic fashion by creating FR and DP hierarchies.

Since while building data and process strategy, independence axiom and decomposition principles are used, discussions about only those aspects have been presented in this chapter. For more discussion on axiomatic design and the information axiom, please refer to Suh (2001).

As you can see in decomposing data and process strategy requirements to identify FR–DP combinations in such a way that either a design is uncoupled or decoupled, the independence axiom is quite useful. The FR–DP combinations corresponding to the top five data and process strategy requirements (in Figure 3.2) are shown in Figure 3.6. Note that, in Figure 3.6, FRs correspond to high-level requirements to support each component of the data and process strategy, while DPs are created to satisfy each FR.

The DPs are aligned with FRs to deliver value to the organization.

Using the decomposition process described above, the top five data and process strategy FR–DP pairs have been decomposed into lower levels. Figure 3.7 shows

FR #	Functional requirements (*what* to accomplish)	DP #	Design parameters (*how* to accomplish)	
FR1	Data valuation	DP1	Valuation quantification model	
FR2	Data innovation	DP2	Discovery tools	2
FR3	Decision support	DP3	Embedded data quality metrics	3
FR4	Risk management and compliance	DP4	Risk mitigation framework	1
FR5	Data access control	DP5	Security framework	5
FR6	Data exchange	DP6	Data delivery	
FR7	Monitoring, controlling, and reporting	DP7	Real-time business insights for making decisions	
FR8	Build-in quality	DP8	Source system certification	4
FR9	Data as services	DP10	Data as service platforms	

FIGURE 3.6 Data and process strategy FRs and DPs.

Decomposed functional requirements (*what* to accomplish)	Decomposed design parameters (*how* to accomplish)
FR81 = Identify authoritative source systems FR82 = Identify and prioritize CDEs at source level FR83 = Determine business thresholds FR84 = Evaluate dimensional level scores FR85 = Proactively monitor and control data at the point of entry FR86 = Certify CDEs	DP81 = Criterion for determining authoritative sources DP82 = CDE rationalization and prioritization tool DP83 = SPC and business SMEs inputs DP84 = Dimensions of interest, business rules, profiling DP85 = Automated correction, cleansing, and standardization to meet requirements DP86 = Mechanism to compare with thresholds and certify

DS8 decoupling - design equation

$$\begin{bmatrix} FR81 \\ FR82 \\ FR83 \\ FR84 \\ FR85 \\ FR86 \end{bmatrix} = \begin{bmatrix} X & 0 & 0 & 0 & 0 & 0 \\ X & X & 0 & 0 & 0 & 0 \\ X & X & X & 0 & 0 & 0 \\ X & X & X & X & 0 & 0 \\ X & X & X & X & X & 0 \\ X & X & X & X & X & X \end{bmatrix} \begin{bmatrix} DP81 \\ DP82 \\ DP83 \\ DP84 \\ DP85 \\ DP86 \end{bmatrix}$$

DS8 decoupling

	Functional requirement	Design parameters					
		DP81 Authoritative sources identification	DP82 CDE identification	DP83 Statistical process control	DP84 Data quality scorecard	DP85 Automated cleansing	DP86 Data quality business rules
FR81	Source system identification	X	0	0	0	0	0
FR82	CDE identification	X	X	0	0	0	0
FR83	Data quality threshholding	X	X	X	0	0	0
FR84	Data quality metrics aggregation	X	X	X	X	0	0
FR85	Incoming data monitoring	X	X	X	X	X	0
FR86	CDE certification	X	X	X	X	X	X

FIGURE 3.7 Decomposition of the FR8–DP8 pair corresponding to *build-in quality.*

the decomposition of the FR–DP pair associated with *build-in quality* and the corresponding decoupled design matrix.

The FR–DP analyses of the other top four FRs are described henceforth.

Functional Requirements–Design Parameters Decomposition—Data Innovation

Functional Requirements

FR21 = Partner with the businesses to understand uses of cross-unit and cross-functional data

FR22 = Provide risk, market, macro, and micro insights

FR23 = Provide intelligence about customers, suppliers, and the network of relationships

FR24 = Optimize business opportunities to improve processes performance and efficiency

FR25 = Validate innovation findings and compare them to best-in-class industry benchmarks

Design Parameters

DP21 = Operating guidelines

DP22 = Risk and market data

DP23 = Appropriate analytics to obtain quality insights and understand relationships

DP24 = Application of analytics and discovered relationships

DP25 = External engagement and partnering model

The corresponding design equation is:

$$\begin{bmatrix} FR21 \\ FR22 \\ FR23 \\ FR24 \\ FR25 \end{bmatrix} = \begin{bmatrix} X & 0 & 0 & 0 & 0 \\ X & X & 0 & 0 & 0 \\ X & X & X & 0 & 0 \\ X & X & X & X & 0 \\ X & X & 0 & X & X \end{bmatrix} \begin{bmatrix} DP21 \\ DP22 \\ DP23 \\ DP24 \\ DP25 \end{bmatrix}$$

Functional Requirements–Design Parameters Decomposition—Decision Support

Functional Requirements

FR31 = Select high-quality data from all data available for the intended business goal(s)
FR32 = Understand uses of data in existing projects
FR33 = Identify opportunities for using internal or external data
FR34 = Classify and prioritize decision needs based on the magnitude of impact (low, medium, high, short-term, long-term)
FR35 = Validate decisions (quality of insights) on a periodic basis and adjust/revise as appropriate

Design Parameters

DP31 = Data asset *marketplace* with embedded quality metrics in each asset
DP32 = Information product overlap matrix/data use and sharing matrix
DP33 = Data acquisition proposal pipeline
DP34 = Data certified as *usable*
DP33 = An expert panel within the company/outside agencies

The corresponding design equation is:

$$\begin{bmatrix} FR31 \\ FR32 \\ FR33 \\ FR34 \\ FR35 \end{bmatrix} = \begin{bmatrix} X & 0 & 0 & 0 & 0 \\ X & X & 0 & 0 & 0 \\ X & X & X & 0 & 0 \\ X & X & X & X & 0 \\ X & X & 0 & X & X \end{bmatrix} \begin{bmatrix} DP31 \\ DP32 \\ DP33 \\ DP34 \\ DP35 \end{bmatrix}$$

Functional Requirements–Design Parameters Decomposition—Data Risk Management and Compliance

Functional Requirements

FR41 = Identify and define threats and risks
FR42 = Assess the likelihood of occurrence and the impact of risks and evaluate existing controls

FR43 = Assess risks and determine responses
FR44 = Address the opportunities identified
FR45 = Develop, test, and implement plans for risk treatment
FR46 = Provide ongoing monitoring and feedback

Design Parameters

DP41 = A consolidated document indicating all threats and risk based on inputs from subject-matter experts (SMEs)
DP42 = Failure mode and effect analysis (FMEA) with SMEs
DP43 = Response prioritization
DP44 = Risk quantification and impact analysis
DP45 = Scenario-based risk mitigation plans
DP46 = Automated monitoring and feedback mechanism

The corresponding design equation is:

$$\begin{bmatrix} FR41 \\ FR42 \\ FR43 \\ FR44 \\ FR45 \\ FR46 \end{bmatrix} = \begin{bmatrix} X & 0 & 0 & 0 & 0 & 0 \\ X & X & 0 & 0 & 0 & 0 \\ X & X & X & 0 & 0 & 0 \\ X & X & X & X & 0 & 0 \\ X & X & X & X & X & 0 \\ X & X & X & X & X & X \end{bmatrix} \begin{bmatrix} DP41 \\ DP42 \\ DP43 \\ DP44 \\ DP45 \\ DP46 \end{bmatrix}$$

Functional Requirements–Design Parameters Decomposition—Data Access Control

Functional Requirements

FR51 = Understand data access needs
FR52 = Request data authorization
FR53 = Approve authorization request(s)
FR54 = Access revocation (short-term/long-term)

Design Parameters

DP51 = Collection of needs by data owner
DP52 = Mechanism for analysts to request data
DP53 = Authorization of data access by data security officers
DP54 = Access review schedule execution

The corresponding design equation is:

$$\begin{bmatrix} FR51 \\ FR52 \\ FR53 \\ FR54 \end{bmatrix} = \begin{bmatrix} X & 0 & 0 & 0 \\ X & X & 0 & 0 \\ X & X & X & 0 \\ X & X & X & X \end{bmatrix} \begin{bmatrix} DP51 \\ DP52 \\ DP53 \\ DP54 \end{bmatrix}$$

		FR2: Data innovation					FR3: Decision support					
		FR21	FR22	FR23	FR24	FR25	FR31	FR32	FR33	FR34	FR35	TOTAL
Discovery tools	DP21	X	0	0	0	0	0	0	0	0	0	1
	DP22	X	X	0	0	0	0	0	0	0	0	2
	DP23	X	X	X	0	0	0	0	0	0	0	3
	DP24	X	X	X	X	0	0	0	0	0	0	4
	DP25	X	X	0	X	X	0	0	0	0	X	5
Embeded DQ metrics	DP31	0	0	0	0	0	X	0	0	0	0	1
	DP32	0	0	0	0	0	X	X	0	0	0	2
	DP33	0	0	0	0	0	X	X	X	0	0	3
	DP34	0	0	0	0	0	X	X	X	X	0	4
	DP35	0	0	0	0	X	X	X	0	X	X	5

FIGURE 3.8 FR–DP Matrix with two FRs and two DPs.

End-to-End Functional Requirements–Design Parameters Matrix

There are several advantages to look at an end-to-end FR–DP matrix corresponding to important FRs. In Figure 3.8, an FR–DP matrix corresponding to two FRs, data innovation (FR2) and decision support (FR3), has been shown.

From Figure 3.8 we can

- Understand complexity of requirements across multiple functions by detecting interdependencies with DPs
- Detect overlap among DPs and in sequencing execution for resource planning
- Understand that focus should be on DP25 and DP35 as they are impacting 5 FRs each. They should be followed by DP24 and DP34 that are impacting 4 FRs each

Thus, axiomatic design principles are extremely useful to prioritize data management requirements and satisfy them in a systematic and scientific fashion.

3.5 METRICS MANAGEMENT[2]

As you can see, in any design activity, it is important to define requirements (*the whats*) and corresponding DPs (*the hows*) in a systematic way so as to avoid complexities at the later stage. Equally important, however, is identifying suitable metrics for measuring FRs and DPs. The accuracy of these metrics is key so as to make sure that desired outcome(s) are achieved. In this section, we will discuss the importance of managing these metrics.

For managing metrics, four steps are usually required.

Step 1: Defining and Prioritizing Strategic Metrics

In this first step, all business units should agree upon the common metrics that are meant to achieve organizational success. After defining these metrics, they need to be prioritized. To do this, we can use a similar approach to that was discussed earlier for prioritizing data and process strategy requirements. Figure 3.9 shows an illustrative example of prioritizing the strategic metrics.

---

[2] The author would like to thank Chris Heien and Javid Shaik for their involvement and help in this effort.

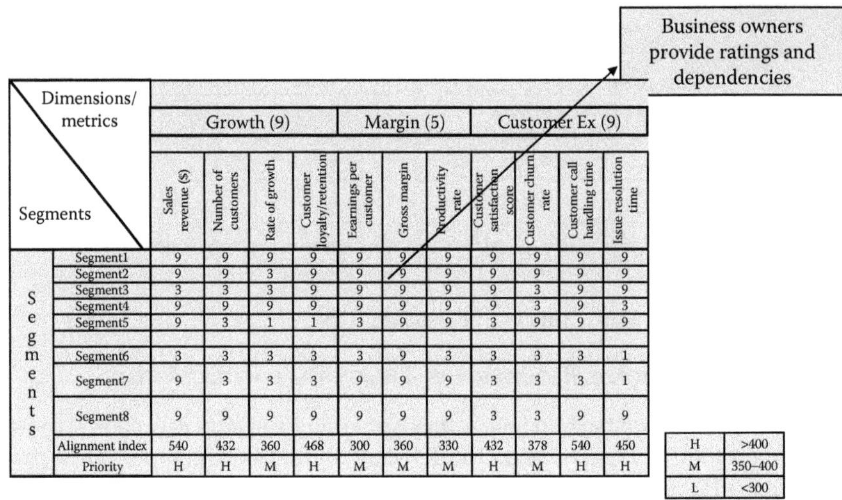

Dimensions/metrics	Growth (9)				Margin (5)			Customer Ex (9)			
Segments	Sales revenue ($)	Number of customers	Rate of growth	Customer loyalty/retention	Earnings per customer	Gross margin	Productivity rate	Customer satisfaction score	Customer churn rate	Customer call handling time	Issue resolution time
Segment1	9	9	9	9	9	9	9	9	9	9	9
Segment2	9	9	3	9	9	9	9	9	9	9	9
Segment3	3	3	3	9	9	9	9	9	3	9	9
Segment4	9	9	9	9	9	9	9	9	3	9	3
Segment5	9	3	1	1	3	9	9	3	9	9	9
Segment6	3	3	3	3	3	9	3	3	3	3	1
Segment7	9	3	3	3	9	9	9	3	3	3	1
Segment8	9	9	9	9	9	9	9	3	3	9	9
Alignment index	540	432	360	468	300	360	330	432	378	540	450
Priority	H	H	M	H	M	M	M	H	M	H	H

H	>400
M	350–400
L	<300

FIGURE 3.9 Prioritizing the strategic metrics (illustrative example).

In this example, strategic metrics were defined for dimensions such as growth, margin, and customers across all segments and product lines. The dimensions are ranked on a scale of 1 to 9 (9 being the most important and 1 being least important) and the metrics are ranked based on their importance to the segments. The weighted score of the dimension scores and metrics importance to the segments (9 being very important and 1 being least important) will provide an alignment index that will help to prioritize the metrics. In this example, a score of 400 and above is considered high priority, a score of 300 to 400 is considered medium priority, and a score of less than 300 is considered low priority.

Step 2: Define Goals for Prioritized Strategic Metrics

The term *goal* here refers to the target level of performance. Goal-setting is to be done by businesses along with other groups (sales, analytics, technology, etc.) for consensus. Historical data analysis and comparison with best-in-class are some of the important activities to consider in setting up goals.

Step 3: Evaluation of Strategic Metrics

For any metric, we should consider the following three things (Figure 3.10): (1) the goal as defined in the previous step; (2) capability, or the actual performance level;

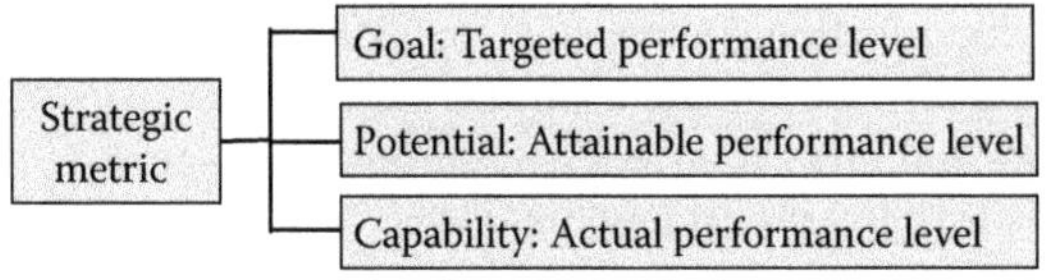

FIGURE 3.10 Measurement aspects of the strategic metrics.

and (3) potential, or the performance level that can be attained after removing special causes of variation.

Note that a process will reach its potential if we remove special causes of variation and potential that may be less than the goal. For reducing the gap between the potential and the goal, significant changes to system are needed (by working on common causes of variation).

Since the process performance is dependent on special causes and common causes of variation and metrics are used to measure the performance, a brief discussion about the causes of variation is presented next.

Common Causes and Special Causes

The interest in understanding the variation due to common causes and special causes is the result of the thinking of Dr. Walter A. Shewhart. According to Dr. Shewhart, variation can come from two types of causes: common causes and special causes. Use of a control chart can help to identify these two kinds of variation. Special causes are also known as assignable causes and common causes are also known as chance causes. Special cause variation occurs when something significant or special happens to the system. Special causes can be detected by out-of-control points or outliers or anomalies on the control chart. Common cause variation occurs because of the inherent nature of the system. They are represented by points on the control charts that are within control limits and that usually stay the same or close within a sample (day to day, shift to shift, or lot to lot) if no change is made to the system. Common cause variation is also known as the expected variation of the process, as it is inherent in the system.

If all points on the control chart corresponding to any process or system are within the control limits (that is, if variation due to assignable causes is not present), then we can say that the process is under the influence of common cause variation and is expected to exhibit the following patterns:

1. Of all the data points, 99.73% of them lie between three standard deviations from the mean. As we know, the statistical control limits are established based on three standard deviations on either side of the mean. The limit on the positive side of the mean is called the upper control limit (UCL) and the limit on the negative side is known as the lower control limit (LCL). Figure 3.11 shows the normal distribution of a variable with one, two, and three standard deviation (σ) distances from the mean. Note that 68.26% of points lie between 1σ limits, 95.46% points lie between 2σ limits, and 99.73% points lie between 3σ limits from the mean. This figure also shows how the process performance compares with upper and lower specification limits. The specifications are thresholds are provided by the customer, supplier, or manufacturer and they can also be based on historical data analysis and subject matter expertise.
2. If the data points exhibit common cause variation, then there should not be any trends or patterns on the control chart. In other words, the data points should be randomly distributed without any specific patterns. There are established rules for evaluating/identifying patterns associated with control charts.

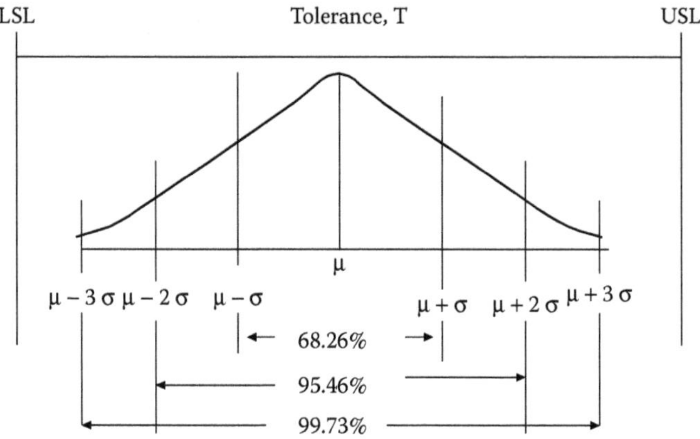

FIGURE 3.11 Distribution of normal population.

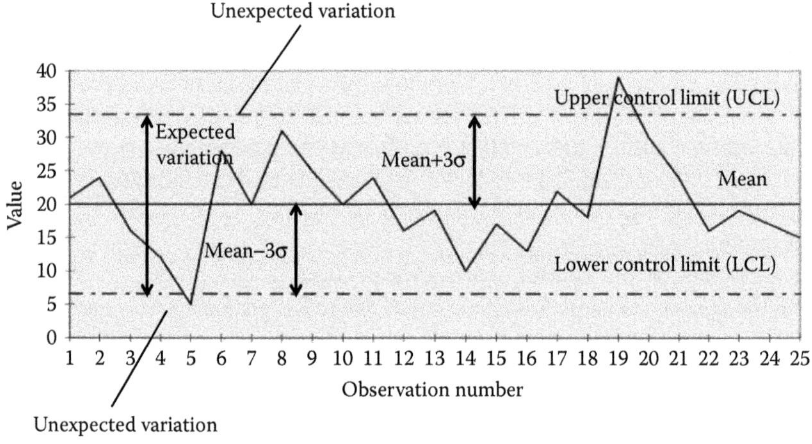

FIGURE 3.12 Various components of a control chart.

If a process exhibits only common cause variation, then it is said to be statistically *in control.* Figure 3.12 describes various components of the control chart that help in finding out if a process is in control or not.

There are seven established conditions with which we can evaluate process conditions (in control or out of control). These are called the Western Electric Run rules and are listed below:

1. The presence of any point outside the control limits
2. The presence of seven consecutive points on the same side of the center line
3. The presence of seven consecutive points increasing or decreasing
4. The presence of two of three points beyond two standard deviations on either side of the mean
5. The presence of four of five points beyond one standard deviation on either side of the mean

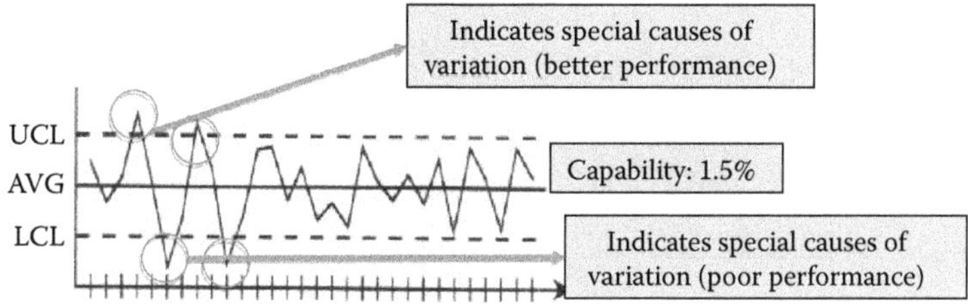

FIGURE 3.13 Control chart and an explanation of capability.

6. The presence of 14 consecutive points alternating up and down
7. The presence of 14 consecutive points on either side of the mean

If the data points satisfy these rules, we can say that the process is out of control or under the influence of special cause variation.

From the above discussion, it should be clear to readers that control charts are an effective way to understand the variability in the system. In metrics evaluation, the control chart is the basis for determining the capability of processes. Figure 3.13 shows an illustrative example with the average (central line) of a growth metric representing the existing capability in the presence of special cause variation. The points above the UCL are good special causes indicating better-than-average performance and the points below the LCL are bad special causes indicating lower-than-average performance.

In order to reach a goal from the capability level, there are two stages that must be completed: (1) eliminate special causes of variation to reach potential, and (2) understand system limitations (chance causes of variation) and act on them to reach goal.

Step 4: Discovery of Root Cause Drivers

This step aims at understanding the chance cause variation and how we can reduce its effect. To reduce the effect of chance cause variation, we need to identify the root cause drivers corresponding to the metrics, and improvement activities should be performed on these drivers to get to the corresponding goals. These root cause drivers are the same as the factors or critical data elements that were discussed in earlier chapters. Figure 3.14 shows the path to reach from the capability to the goal for the growth metric that is being considered. It is important to conduct a likelihood analysis of going from the potential to the target so that we can determine if it is worth undertaking the effort of

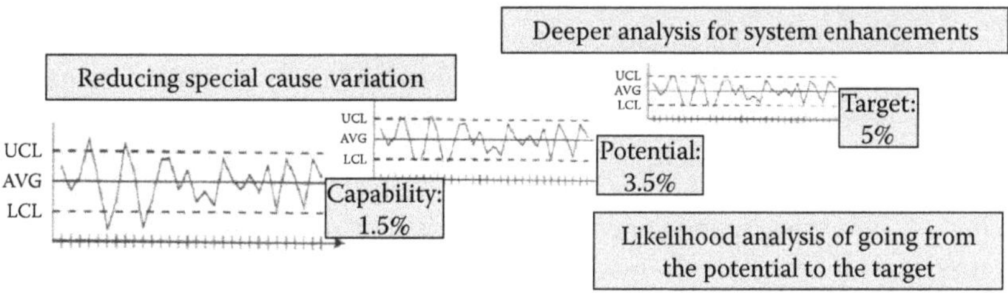

FIGURE 3.14 Capability to goal using control charts.

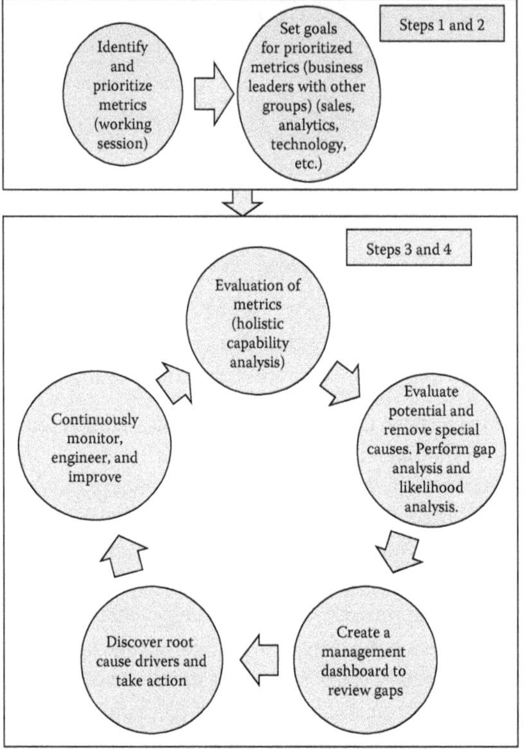

FIGURE 3.15 Performing metrics management.

going from the potential to the target with the existing system, or we need to do a deeper analysis to identify and make system enhancements to address root cause drivers.

Figure 3.15 shows how we can perform all of the four activities of the metrics management in the form of a continuous improvement cycle. The following are important items to remember in order to carry out this cycle:

- A working session is needed to prioritize metrics and set goals based on the dimensions and business areas (steps 1 and 2).
- A systematic approach is required to perform holistic capability analysis at the metrics level and to identify root cause drivers impacting gap between potential and target (steps 3 and 4).
- A team of analytical engineers is needed to conduct gap analysis, a likelihood analysis of going from the potential to the target, to create a management reporting system, to discover root cause drivers, and to continuously improve the performance.

Having of a metrics management framework will also help in evaluating the effectiveness of investments made on analytical techniques including artificial intelligence and machine learning techniques by looking at the gains achieved in terms of reaching the target.

4 Robust Quality—An Integrated Approach for Ensuring Overall Quality

4.1 INTRODUCTION

As mentioned in Chapter 2, there is a need to look at the quality approach holistically by combining aspects of quality in data science and process engineering. The integrated approach should attempt to satisfy overall quality requirements to achieve robust quality. Many times, data quality (DQ) and process quality initiatives are operated with a silo mentality. The proposed integrated approach will help to combine these two aspects of quality and help to quickly solve business problems more accurately. This chapter discusses the development of such an integrated approach, which will aim at improvements and stable operations. The integrated approach is described by considering the DQ approach, lean Six Sigma methods, and Taguchi methods (developed by Genichi Taguchi) for quality engineering.

4.2 DEFINE, MEASURE, ANALYZE, IMPROVE, AND CONTROL-BASED INTEGRATED APPROACH FOR ROBUST QUALITY

In any Six Sigma problem-solving approach, it is important to look at quality holistically. This will help organizations to solve related issues more efficiently and effectively so as to make better business decisions. In this section, an approach is provided to address robust quality through the Define, Measure, Analyze, Improve, and Control (DMAIC) phases. Figure 4.1 shows a robust quality approach based on DMAIC. The descriptions of activities in these phases are presented in the following sections.

The Define Phase

This phase focuses on project definition by establishing the scope, objectives, and resources needed as well as project plans with strong governance and stakeholder support.

The project scope can usually be established based on issues and targeting the business processes causing the issues. The scope should focus on understanding the critical processes, associated metrics, and control and monitoring activities.

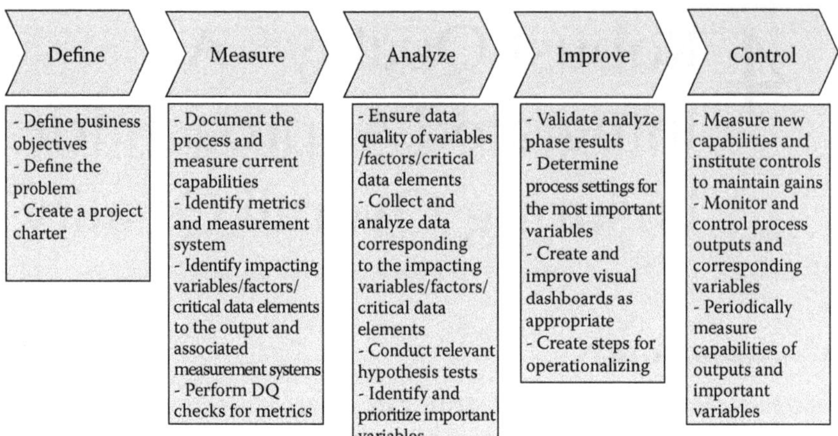

FIGURE 4.1 DMAIC approach for robust quality.

As mentioned earlier in Chapter 2, the most important activity of the Define phase is the creation of a project charter by establishing the scope, objectives, resources, expected business value, and role clarity. It should also outline broad time frames and core deliverables. A typical project charter should have the elements shown in Table 4.1.

TABLE 4.1
Project Charter

Charter Element	Description	Comments		
Project Goal	• Should be a clear goal statement with a success measure • Goal should be Specific, Measurable, Attainable, Relevant, and Time-based			
Scope	• Project scope should include processes, tools, systems, and data that will be used			
Metrics	• Primary metric:			
Resources/Roles	• Estimates of resources and associated roles			
		Phase	**Major Deliverables**	**Date**
Timeline	• Project major deliverables with various phases and timelines	Define		dd/mm
		Measure		dd/mm
		Analyze		dd/mm
		Improve		dd/mm
		Control		dd/mm

Establish role clarity: In the Define phase, the team must clarify the roles related to the project and identify all responsible parties.

Create project plan: Project managers need to prepare a detailed project plan for the five phases of the DMAIC approach with all of the relevant tasks and associated deliverables included.

The Measure Phase

The first important activity in the Measure phase is to document the process and measure the current capability of the associated metrics. Documentation of the process is very important, as it is helpful to look at all activities and also to identify value-added activities and non-value-added activities. A process represents a set of the activities required to develop an input into an output so that customers can use the latter. Usually, a suppliers, inputs, processes, outputs, and customers (SIPOC) analysis, as shown in Figure 4.2, is useful to capture all of the elements of a process.

After defining the process and its elements, the next step is to identify the metrics and associated measurement systems. The discussion on metrics management in Chapter 3 is relevant here. As mentioned in Chapter 3, each metric should have three aspects—specifically, goal, potential, and capability—as part of their measurement system, as shown in the following.

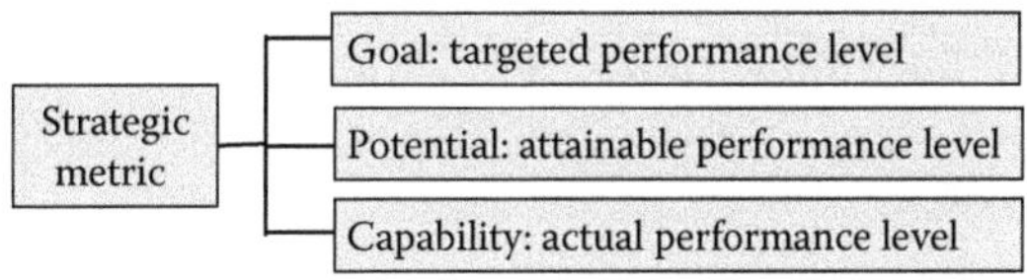

To ensure the DQ of metrics, performing a measurement system analysis (MSA) is quite important. MSA is a method for evaluating how much variation is present in the process of data collection. It helps with understanding the accuracy and reliability of the process of data collection. MSA also helps in evaluating the effectiveness of the data collection plan. After conducting MSA, we can perform other DQ checks as needed with the methods outlined in Chapter 2. After ensuring that metrics are accurate, reliable, and fit for the purpose of usage, it is important to identify variables/factors/critical data elements that impact the output and associated measurement systems. This step will lead to the next phase, which is the Analyze phase.

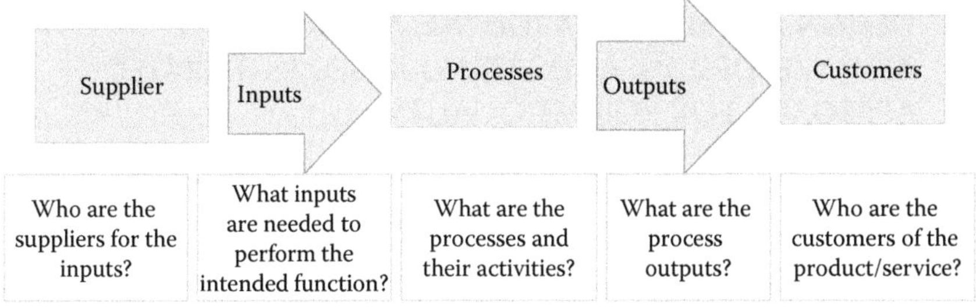

FIGURE 4.2 SIPOC representation to define the process.

The Analyze Phase

We can begin this phase with the consideration of an important set of impacting variables/factors/critical data elements. It is essential to collect data associated with these variables to conduct hypothesis tests in order to prove or disprove hypotheses that these variables are impacting. Prior to collecting data, we need to ensure that we have a reliable measurement system to collect the data and that the DQ levels of these variables are satisfactory. For DQ checks, we can use the method described in Chapter 2. After collecting clean data, we can perform hypothesis tests. At the end of this phase, we should be in a position to isolate the most important variables contributing to the problem and identify the improvement actions that are required to address them.

The Improve Phase

The Improve phase of the DMAIC approach for robust quality focuses on validating the Analyze phase results and determining the right settings for the most important variables. The experimental design approach discussed in Chapter 2 often times will be very useful in determining suitable settings. The outcomes should be practical and should also be analytically sound. This phase also focuses on the creation of visual displays in the form of heat maps, dashboards, and scorecards. At the end of this phase, we should have a clear plan for operationalizing new settings with the appropriate steps determined.

The Control Phase

The Control phase should focus on measuring new capabilities and instituting controls to maintain gains obtained in the Improve phase. In this phase, monitoring and controlling output metrics and corresponding impacting variables usually done through the use of heat maps, scorecards, and dashboards. The use of statistical process control charts is highly recommended in the Control phase. Also, in the Control phase, we should have a plan to periodically measure the capabilities of outputs and important variables.

Figure 4.3 shows the tools/techniques that might be useful in performing various activities to achieve robust quality through DMAIC. References for these tool/techniques can be easily obtained through available literature.

4.3 DESIGN FOR SIX SIGMA (DEFINE, MEASURE, ANALYZE, DESIGN, AND VERIFY)-BASED INTEGRATED APPROACH FOR ROBUST QUALITY

As in the DMAIC robust quality approach, quality should be viewed holistically when designing new products/services using the Design for Six Sigma (DFSS) approach. This will help organizations to develop high quality products/services, which will in turn make them compete and potentially even thrive in the global markets. In this section, an approach is provided to address robust quality through DFSS or Define, Measure, Analyze, Design, and Verify (DMADV). We use the

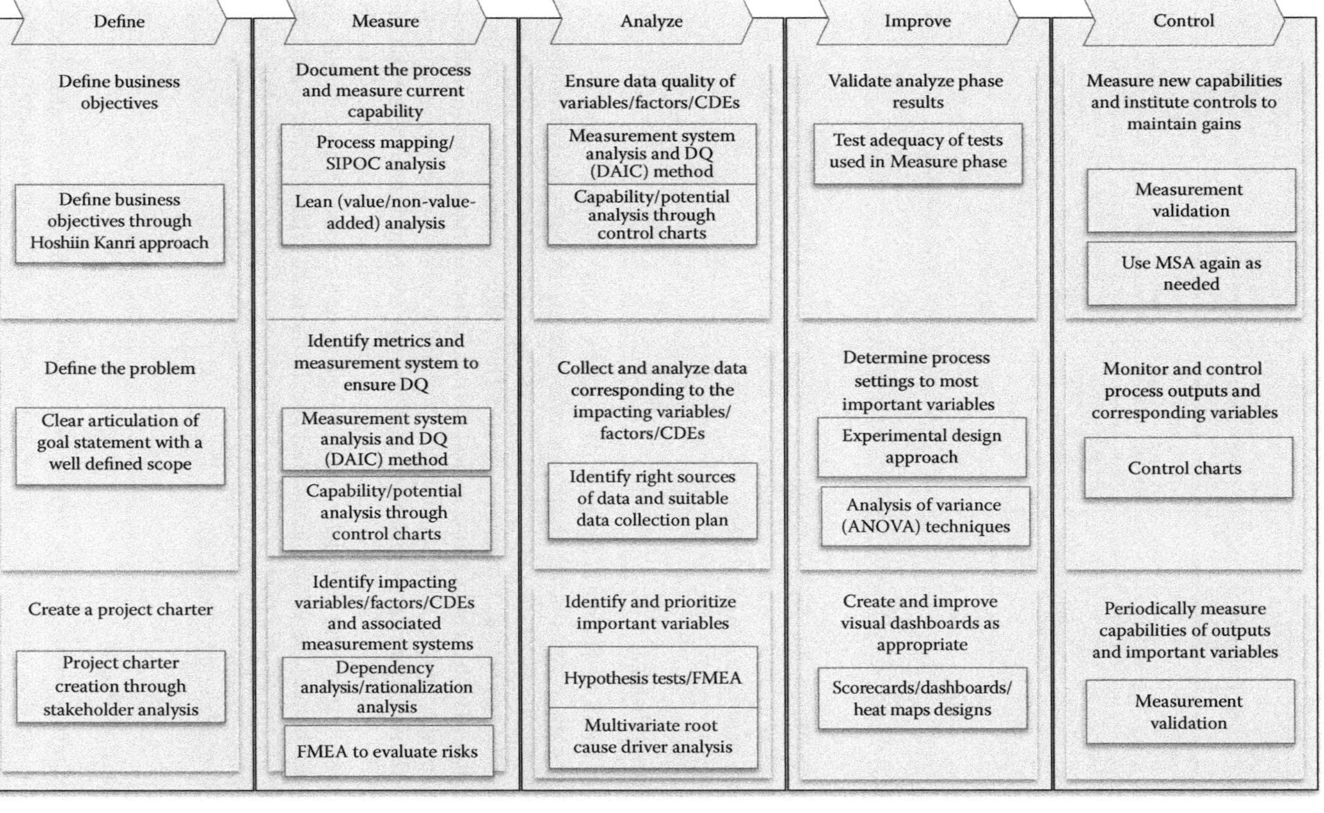

FIGURE 4.3 DMAIC-based robust quality approach (with relevant tools/techniques).

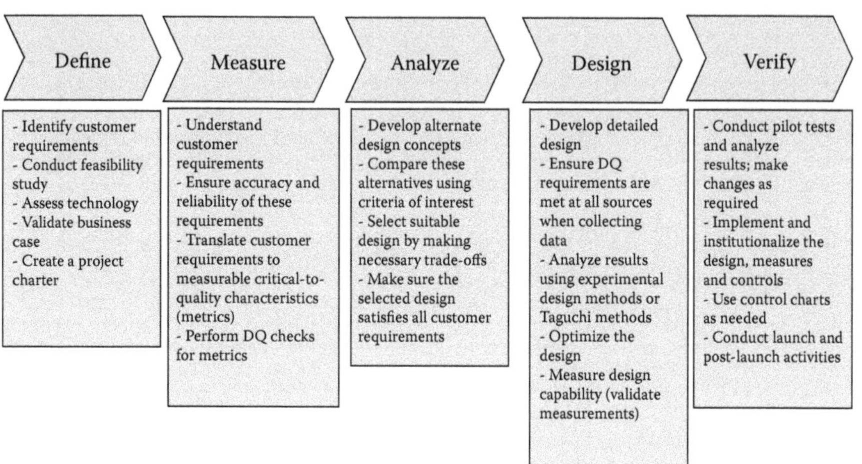

FIGURE 4.4 DFSS or DMADV approach for robust quality.

Design for Lean Six Sigma framework described in Chapter 2 for illustrating various phases, as it covers lean aspects as well. Figure 4.4 shows a robust quality approach based on DMADV steps.

The Define Phase

Since, in DFSS, a successful design is the primary focus, this phase aims to discern customer requirements. After identifying customer requirements, it is recommended that a feasibility study be conducted by assessing existing technology. Then, after this step, the business case must be validated. Here, also, an important activity of the Define phase is the creation of the project charter. The project team is responsible for establishing the scope, understanding the expected business value, and creating the project charter. The charter (as shown in Table 4.1) needs to frame the project and provide the objectives, scope, key participants, and resources. It should also outline broad time frames and core deliverables for all design-related DFSS activities. As mentioned before, the project team must clarify the roles related to the project and identify all responsible parties. Project managers need to prepare a detailed project plan for the five phases of the DFSS/DMADV approach with the relevant tasks and associated deliverables included.

The Measure Phase

In the Measure phase, the first activity is to understand the customer requirements. Then, we need to ensure the accuracy and reliability of the information regarding these requirements. The MSA plays an important role here in ensuring the reliability of the information. After performing MSA, we can perform other DQ checks as needed.

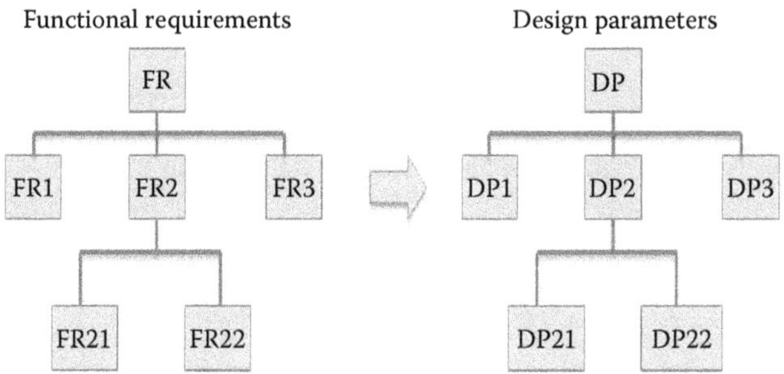

FIGURE 4.5 Decomposition of functional requirements and design parameters.

In the next step, the customer requirements (i.e., the what parts) are translated into measurable quality characteristics (i.e., the how parts). The decomposition approach based on axiomatic design theory is highly recommended for this, although other techniques such as quality function deployment (QFD) can also be used. Figure 4.5 shows the decompositions approach. The high-level requirements (functional requirements) are decomposed to lower-level requirements by identifying corresponding design parameters. In Figure 4.5, the functional requirements represent the *what part* of the equation and the design parameters represent the *how part.*

The Analyze Phase

In the Analyze phase, the first task is to develop alternative design concepts, keeping in mind the requirements obtained from the customers and associated critical quality characteristics. Using the criteria of interest (e.g., costs, usability, customer preferences), we need to compare design alternatives and choose the best one. Pugh Concept Selection is a very widely used method for selecting the best design among various alternatives by making necessary trade-offs. The conceptual design that is selected should ideally be able to satisfy all customer requirements.

The Design Phase

In the Design phase, the conceptual design will take its shape. When the detailed design is developed, it is important to ensure DQ requirements are met at all levels when collecting data. The use of DQ checks is important in validating the data. The experimental design approach and/or Taguchi methods are usually quite important in the Design phase. They help in conducting various experiments and in analyzing the results. By analyzing the results, through design optimization, we can select the factors and corresponding settings that are best-suited to satisfy customer requirements. The last step in this phase is measuring the optimal design capability. Of course, we need to ensure that the measurements are accurate and reliable while measuring design capability. At the end of this phase, we should have a clear plan for operationalizing the design with appropriate steps.

The Verify Phase

In this phase, we must conduct pilot tests as needed and analyze the results to make sure that the new design's capability meets the requirements/standards. In some cases, changes may be required in order to reach the optimal level. This is sometimes called design tuning. After this step, the design should be implementable: to achieve this, we need to institutionalize the new design along with new measures and controls. Statistical process control charts are quite useful in managing measures and controls. In the Verify phase, we should also have a plan to periodically measure the capabilities of output metrics and of important variables.

Figure 4.6 shows the tools/techniques that might be useful in performing various DMADV/DFSS activities in attempting to achieve robust quality. References for these tools/techniques can be obtained through available literature.

4.4 TAGUCHI-BASED INTEGRATED APPROACH FOR ROBUST QUALITY

While designing a new product, Taguchi methods consider the three stages of concept design, parameter design, and tolerance design. However, although there are three stages, most of the Taguchi applications focus on the latter two (parameter design and tolerance design).

Importantly, if you start the designing process with a robust concept selected in the concept stage, the gains in terms of quality will be higher. Techniques like the theory of inventive problem-solving (TRIZ) as well as axiomatic design and P-diagram strategies for robust quality engineering (Jugulum and Frey, 2007) can be used to achieve robustness at the concept level.

As mentioned earlier, the methods of Taguchi's quality engineering are developed based on the following principles. These principles are presented in Chapter 2 and again in the following:

1. Evaluation of the functional quality through energy transformation
2. Comprehension of the interactions between control and noise factors
3. Use of orthogonal arrays (OAs) for conducting experiments
4. Use of signal-to-noise (S/N) ratios to measure performance
5. Execution of two-step optimization
6. Establishment of a tolerance design for setting up tolerances

If we look at DMAIC and DFSS approaches, Taguchi parameter design methods will play important roles in the Analyze, Improve, and Design phases. A typical parameter design approach has the following three stages:

1. Defining stage
2. Planning stage
3. Execution stage

Figure 4.7 summarizes activities (including DQ-related) in these stages.

The descriptions of activities in these stages are as follows henceforth.

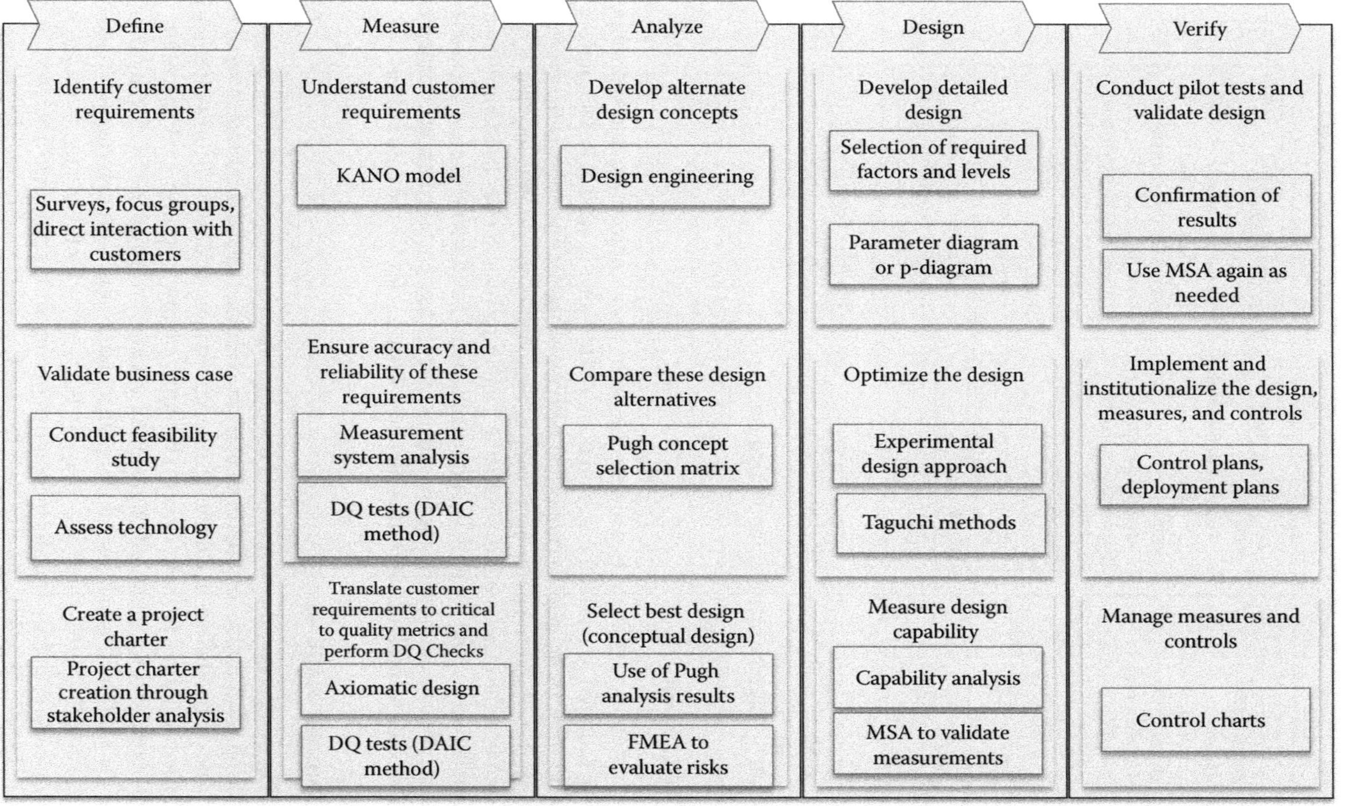

FIGURE 4.6 DMADV-based robust quality approach (with relevant tools/techniques).

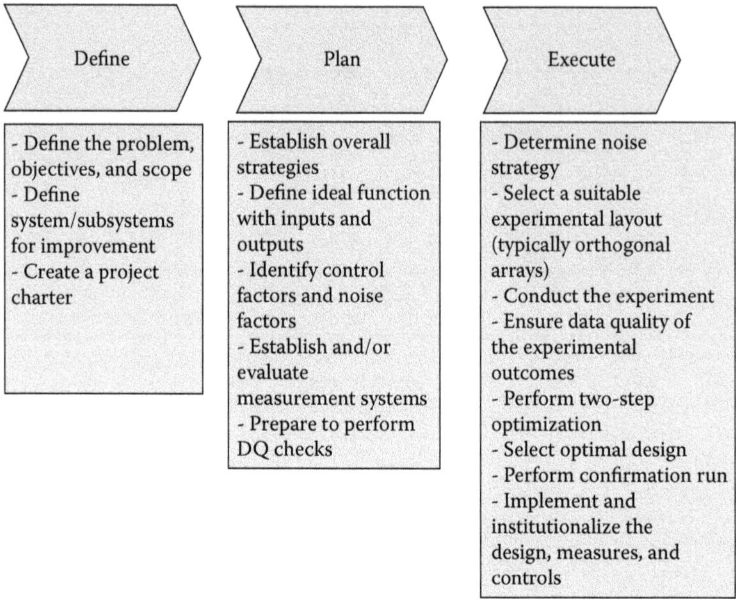

FIGURE 4.7 Taguchi approach for robust quality.

The Define Stage

The first step of this stage is defining the problem, its objectives, and the scope. It is also required to define the system or subsystem that is subject to the improvement activities. Here, also, an important activity of the Define stage is the creation of the project charter. The project team is responsible for establishing the scope, understanding the expected business value, clarifying related roles by identifying all responsible parties, and creating the project charter. The charter (Figure 4.1) needs to frame the project and provide the objectives, scope, key participants, and resources as well as outline broad time frames and core deliverables for the entire effort.

The Planning Stage

In the Planning stage, the first item is establishing overall strategies. This includes considering the impacts of improvements on cost, market, and customer satisfaction. As described before in Chapter 2, Taguchi parameter design focuses on energy transformation—that is, how effectively input energy is transmitted to obtain a desired output. As we know, all engineering systems are governed by a relationship (called the ideal function) between the input and the output through energy transformation. The S/N ratio (which is defined as the ratio of useful energy and wasteful energy) measures energy transformation that occurs within a design. Therefore, it is critically important to identify the ideal function, as the Taguchi approach uses this relation to bring the actual system closer to the ideal state (Figure 4.8) through design optimization. A p-diagram representation is also very important in parameter design,

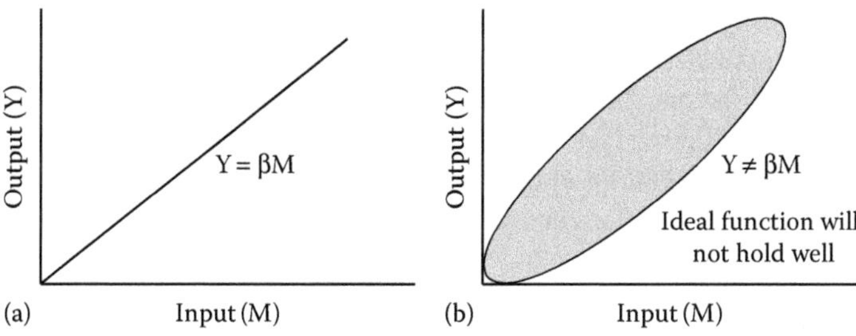

FIGURE 4.8 (a) Ideal function and (b) actual function (reality).

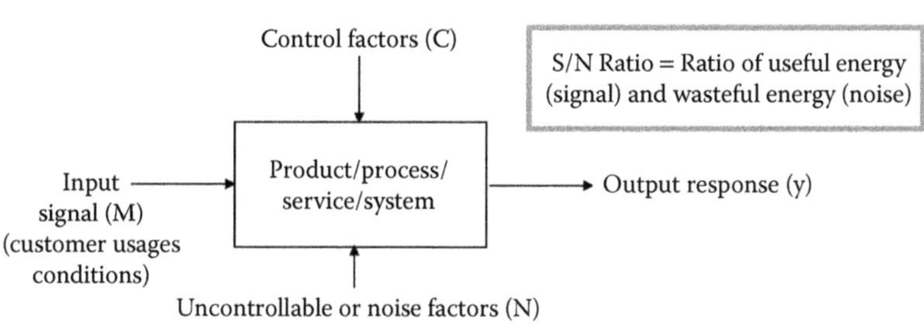

FIGURE 4.9 p-diagram representation.

as it captures all of the elements associated with the product/system. A p-diagram representation is shown in Figure 4.9.

Here, also, it is quite important to establish and/or evaluate measurement systems to measure the energy transformation and corresponding outputs. We should also have plans to perform tests and ensure the quality of test data with suitable DQ approaches, as outlined in Chapter 2.

The Execute Stage

The first step in the Execute stage is to determine noise strategy. Noise strategy refers to the treatment of noise or uncontrollable factors. Usually, the energy transformation is affected by the adverse effects of noise factors. Taguchi's approach aims at the design of a product/process that is insensitive to noise factors. Although there are different noise strategies, Taguchi recommends compounding noise strategy. Using compounding strategy, different noise factors are compounded into one overall noise factor with two or three levels by considering extreme conditions. So, through this strategy, for experiment, there will be one compounded noise factor irrespective of the multiple noise factors that were considered at the beginning of the experiment. Compounding strategy also helps to reduce the resources and time needed in conducting the experiment, since we are reducing the levels associated with noise factors.

After identifying the noise factors and a noise strategy, the next step is to select a set of control factors and a suitable experimental layout that can accommodate control factors and noise factors. In the Taguchi methods, typically, orthogonal arrays are chosen for accommodating factors for the experiment. As mentioned earlier, orthogonal arrays are fractional factorial designs wherein we use only a fraction of the total number of experiments. After conducting experiments based on selected orthogonal array layout, the next step will be to focus on recording the experimental results.

Note that it is very important to ensure DQ requirements for all measurements at the end of each experiment. This includes validating measurement systems and ensuring reliability of noise factor measurements.

Following their collection, the experimental results are analyzed through S/N ratios to find out the optimal design with impacting factors and levels. The optimal design is selected through two-step optimization, as follows.

Step 1: In this step, the combination that maximizes the S/N ratio is identified. The maximum S/N ratio means minimum variation and more consistent performance.

Step 2: In this second step, the factors are adjusted to get the desired performance.

After identifying the optimal design, a confirmation experiment is conducted to make sure that the results are repeatable. At this point, the necessary adjustments need to be made to the design (via design tuning) to get the desired performance levels. If the confirmation run and associated results are satisfactory, then the design can be implemented into actual production. At this point, we will need to institutionalize the new design along with new measures and controls. As mentioned earlier, statistical process control charts are quite useful in managing measures and controls. We should also periodically measure S/N ratios as well as the capabilities of output metrics and of important variables to ensure reliability.

4.5 MEASURING ROBUST QUALITY

So far, in this chapter, the discussions have been focused on the importance of integrating data quality and process quality to achieve robust quality with different approaches. Now, in this section, we will discuss how we can measure robust quality by using the idea of robust quality index (RQI), which takes into account both process quality and DQ aspects. First, we will describe the approach at the metric level and then how it could be rolled up to the enterprise level. Consider two metrics (for example, understanding of customer concerns and customer issue resolution time). For these two metrics, we need to examine the DQ levels by defining suitable DQ dimensions. Table 4.2 shows an illustration with the DQ dimensions along with corresponding scores and overall DQ scores.

After obtaining DQ scores for the metrics, in the next step, we need to examine the process quality aspect of the metrics. The first metric (understanding of customer concern) can be measured on a scale of 1 to 10, with 10 being the highest score

TABLE 4.2
DQ Evaluation of Two Metrics

DQ Dimension	Completeness	Conformity	Validity	Accuracy	DQ Score
Metric 1	95%	90%	85%	78%	87%
Metric 2	96%	93%	88%	87%	91%

DQ: Data quality.

and 1 being the lowest score, and the second metric (issue resolution time) can be measured in minutes or another unit of time. After this, we need to determine the target performance of the metric. For measuring RQI, it is required that deviations of DQ scores and process quality scores from the target be calculated. Since DQ scores are measured on a scale of 0 to 100, it would be better to transform process quality scores to the scale of 0 to 100. After doing this and defining the targets, the deviations will be calculated and, from the deviations, S/N ratios can be calculated. S/N ratios are measured in decibel (dB) units. Based on the S/N ratios, RQI values are able to be measured on a scale of 0 to 100. Appendix III shows the DQI scores and corresponding mean square deviations and S/N ratios. Readers can reach out to the authors for the methodology details if desired. Table 4.3 shows the details of RQI calculations for these two metrics. Also, note that RQI value can be related to Six Sigma defect levels.

TABLE 4.3
RQI Values for the Two Metrics

Metric 1	Score	Target	Squared Deviation
Metric 1-DQ	87	90	9
Metric 1-PQ	90	100	100
		MSD	7.38
		SNR (log)	−8.68
		RQI	56.59
Metric 2	**Score**	**Target**	**Squared Deviation**
Metric 2-DQ	91	95	16
Metric 2-PQ	85	100	225
		MSD	10.98
		SNR (log)	−10.40
		RQI	47.98

MSD: Mean square deviation; SNR: S/N ratio; DQ: Data quality; PQ: Process quality; and RQI: Robust quality index.

TABLE 4.4
RQI for the Function- or Business-Unit-Level

Metric	Score	Target	Squared Deviation
Metric 1-DQ	87	90	9
Metric 2-DQ	91	95	16
Metric 1-PQ	90	100	100
Metric 2-PQ	85	100	225
		MSD	9.35
		SNR	−9.71
		RQI	51.45

MSD: Mean square deviation; SNR: S/N ratio; DQ: Data quality; PQ: Process quality; and RQI: Robust quality index.

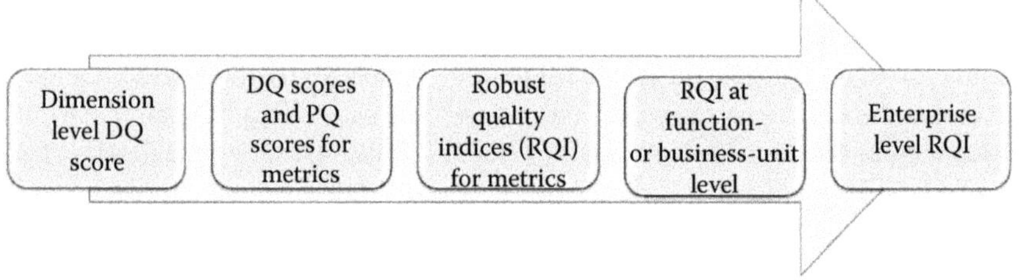

FIGURE 4.10 Getting to enterprise-level RQI.

If these two are the main metrics in a function or business unit of an organization, then they can be rolled up to calculate the function- or business-unit-level RQI. Table 4.4 illustrates this.

Similarly, the function- or business-level-metrics can be rolled up to obtain the enterprise-level RQI. Figure 4.10 illustrates this roll-up process.

5 Robust Quality for Analytics

5.1 INTRODUCTION

In this data-driven world, there is no need for further emphasis on the fact that data capability and analytics capability are important. The quality of analytics (i.e., the quality of the insights obtained through analytics) is as important as the quality of data. In order to obtain accurate and quick analytical results, there should be a structured approach to perform analytics so that they can be scaled to get repeatable and consistent results. In this chapter, we will discuss the importance of having such a process to ensure both data quality (DQ) and analytics quality levels are met in order to enable one to perform high-quality analytics and make appropriate decisions.

5.2 ANALYTICS REQUIREMENTS

The most important requirement for successful analytics execution is support from senior management. Six Sigma programs in companies like General Electric (Boston, MA, USA) and the Bank of America (Charlotte, NC, USA) were successful because of strong senior management support. For executing analytics successfully, a similar kind of support is absolutely critical. When executing analytics across any organization, there should a clear *analytics vision*, followed by a sound *analytics strategy*, as shown in Figure 5.1. The vision includes what will be accomplished through analytics and the strategy should outline various aspects as to how we can accomplish these items. After defining the vision and outlining a suitable strategy, we should be looking at acquiring the analytical talent and resources for the successful deployment of analytics strategy.

After looking at talent and resources, the next step is to make sure that there exists the right kind of data capabilities and analytics platform, as companies often need to deal with huge datasets including those that are unstructured. Having a suitable analytics platform with relevant features such as reporting and monitoring is very essential in order to be able to process the data. Since companies have large amounts of data in various areas it is important for organizations to have a standardized analytics approach with an appropriate operating model for successful execution. Standardization helps emphasize the need for a disciplined approach to execute analytics and helps in choosing the appropriate type of analytics in a given scenario. The operating model will serve as reference for an approach to execute analytics with descriptions on different types of analytics, how to deploy them, how to interpret the results, and associated methodologies. After satisfying analytics requirements, the next stage is to define a process for analytics execution.

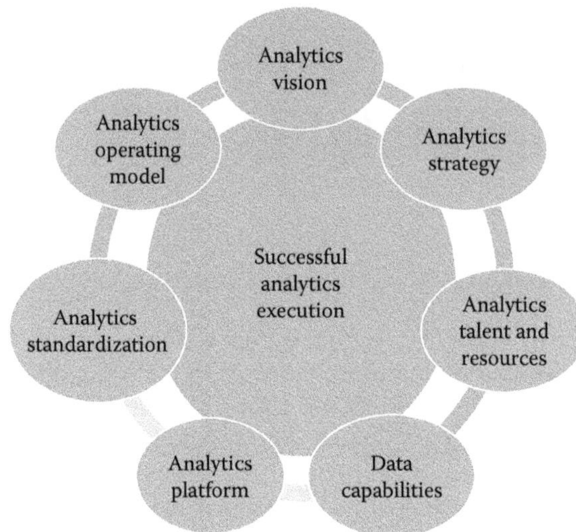

FIGURE 5.1 Successful analytics execution.

5.3 PROCESS OF EXECUTING ANALYTICS

Figure 5.2 outlines an eight-step process for analytics execution for a given problem. The first step in the execution process is to define the problem by understanding the clear purpose. The second and third steps will focus on collecting relevant data and ensuring high DQ. After ensuring high DQ, in the fourth step, the emphasis will be on deciding on the type of analytics we need to use (Table 5.1). The fifth and sixth steps will focus on analytical framework- or model-building and use. The seventh step is required to compare model outputs and actual outcome. For a good model, the gap between these two should be small. If the model is providing intended results (i.e., if the gap is small) then, in the next step, we should focus on the deployment of models outputs in the decision-making activities. If the gap between the model results and actual outcomes is large, then we need to go back to the fifth step to refine the model and follow the subsequent steps again.

Note that the approach outlined in Figure 5.2 is for one problem. Importantly, the same approach should be used for other problems, and we need to standardize the method of deploying the insights (model outputs) for these problems as much as possible.

Note also that, in this approach, step three, *perform DQ checks and ensure DQ*, is quite important because high-quality data are very critical for the successful execution of analytics to obtain meaningful business outcomes. If the quality of the data is not satisfactory, then the analytics results are not going to be valid and this can result in poor decisions and that in turn might result in a huge loss to society. Ronald A. Fisher, a famous statistician, highlighted the importance of the cross-examination of data for meaningful analysis of the data and interpretation of results much before the rapid growth of the data field (Rao, 1997). Cross-examination of data is the same as performing checks to ensure DQ.

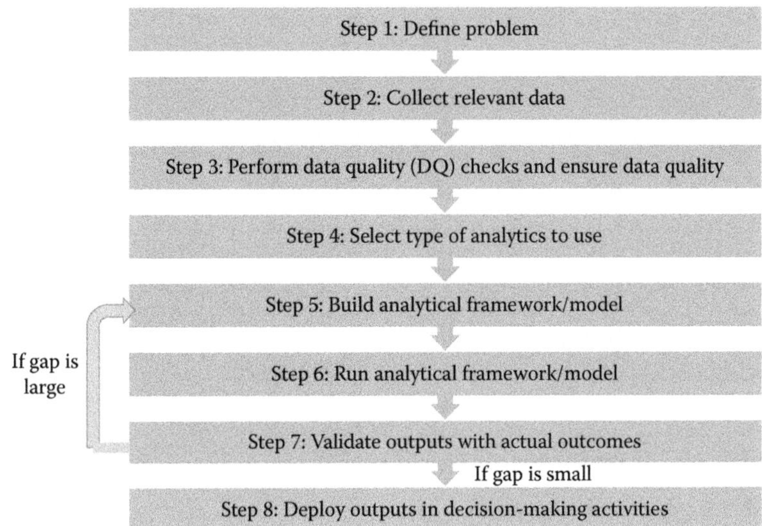

FIGURE 5.2 Analytics execution. (This was published in Jugulum (2014). The author thanks Wiley for granting permission.)

TABLE 5.1
Tools and Techniques for Analytics

Type of Analytics	Tools/Techniques
Preparatory analytics	Simple profiling with DQ rules, SPC charts, and DQ funnel approach, etc.
Descriptive analytics	Data profiling, data mining, descriptive statistics, and capability analysis, etc.
Diagnostic analytics	SPC charts, analysis of variance, correlation and association analysis, and hypothesis tests, etc.
Cause-related analytics	Cause-and-effect analysis, failure mode effects analysis, and parameter diagram (p-diagram), etc.
Predictive analytics	Generalized linear models and regression analysis, etc.
Prescriptive analytics	Experimental design and simulation-based scenario planning, etc.
Reliability-based analytics	Signal-to-noise ratio analysis, failure analysis, and confidence intervals, etc.

The fourth step of the analytics execution process highlights the importance of choosing a right type of analytics. There are different types of analytics that can be selected depending on the problem being dealt with. Figure 5.3 shows analytics taxonomy, along with their intended purposes. They include the types that Gartner (2012) proposed and the types that Jugulum (2014) added to this list. Table 5.1 shows the tools and techniques required for these analytics.

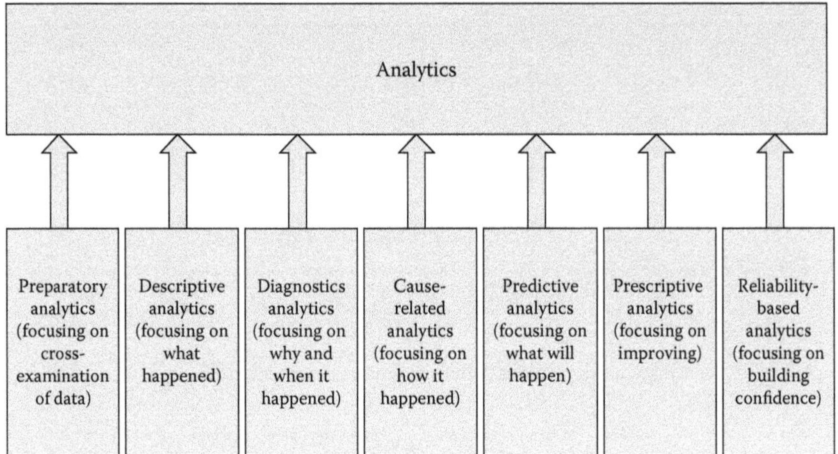

FIGURE 5.3 Taxonomy of analytics.

Brief descriptions of the types of analytics shown in Table 5.1 are provided in the following.

Preparatory analytics: The purpose of preparatory analytics is to conduct cross-examination of data to ensure DQ. Typically, simple data profiling techniques, the DQ funnel approach, and statistical process control (SPC) methods are used in this type of analytics.

Descriptive analytics: The descriptive analytics concept deals with answering questions like what happened to the performance of a particular critical data element (CDE)/factor, system, process, or product. Data profiling and data mining tools as well as stability and capability analysis using descriptive statistics are quite useful in this type of analytics.

Diagnostic analytics: This type of analytics is useful to understand when, where, why, and how a particular problem has occurred. Typical tools and techniques used in this type of analytics include correlation analysis, association analysis, hypothesis tests, analysis of variance, and control charts.

Cause-related analytics: Cause-related analytics are intended to conduct root cause analysis for understanding the causes of problems or failures. The Ishikawa or cause-and effect-diagram, cause and effect matrix, parameter diagram (p-diagram), and failure mode effects analysis are all quite useful for this type of analytics.

Predictive analytics: If the goal is to predict the behavioral patterns associated with a process, product, system, or CDE, then predictive analytics are used. Tool and techniques such as generalized linear models and regression analysis are useful in performing predictive analytics.

Prescriptive analytics: Prescriptive analytics focus on answering questions related to performance improvement. They are useful to answer questions such as how one can improve process performance by taking some actions.

Experimental design approach is very useful here. Simulation-based scenario analysis is also used in prescriptive analytics.

Reliability-based analytics: To estimate the reliability of a product or process or systems or set of models with a required degree of confidence, we use this type of analytics. Failure analysis, confidence intervals, and signal-to-noise ratios are quite useful in this type of analytics.

5.4 ANALYTICS EXECUTION PROCESS IN THE DEFINE, MEASURE, ANALYZE, IMPROVE, AND CONTROL PHASES

The goal of data analytics is to aid organizations in processing large amounts of data so as to make better business decisions that will lead to better outcomes and increase the organization's competitive position.

The processing of data generally requires high-end technologies and good resources with the ability to perform analytics quickly and efficiently. In addition, there is a great need to have a structured approach to perform analytics. The Define, Measure, Analyze, Improve, and Control (DMAIC) framework is also well-suited for executing analytics processes. Figure 5.4 explains the analytics process execution by following the DMAIC steps.

THE DEFINE PHASE

The Define phase always focuses on project definition by establishing the scope, objectives, and resources needed as well as project plans with strong governance and stakeholder support. The project scope can be established by selecting the targeted business processes. In the Define phase, the team must clarify the roles related to the project and identify all responsible individuals. As we know, the most important activity in the Define phase is the creation of the project charter that establishes the scope, objectives, resources, expected business value, and role clarity. The project managers need to prepare a detailed project plan for the five phases of DMAIC with the tasks and associated deliverables included. A discussion on the project charter along with its important elements is provided in Chapter 4.

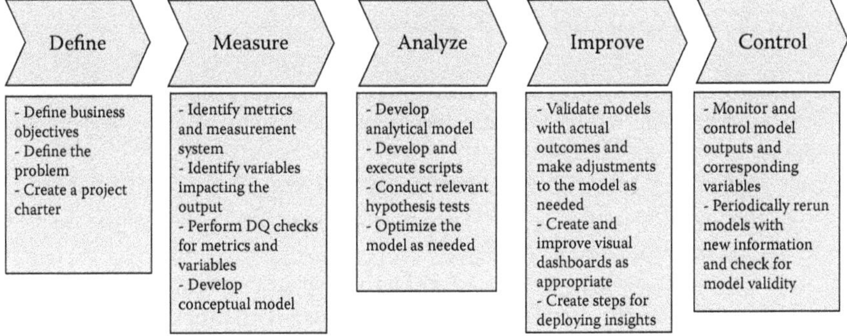

FIGURE 5.4 Structured DMAIC approach for analytics execution.

The Measure Phase

In the Measure phase, the first step is to identify the metrics and associated measurement systems. The discussion on metrics management in Chapter 3 is very useful for the deliverables of this phase. As we discussed earlier, each metric should have three aspects (goal, potential, and capability) as part of their measurement system, as shown in the following.

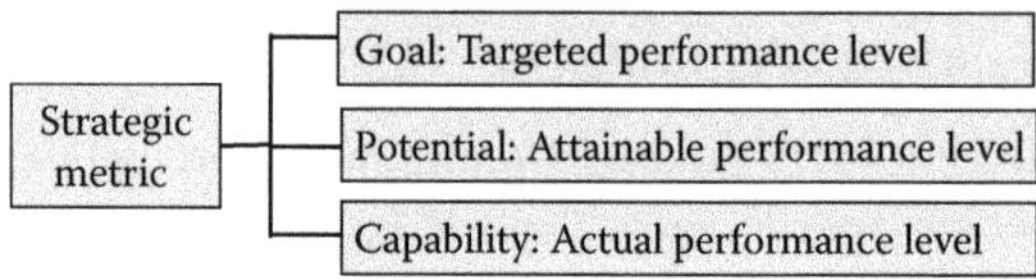

To obtain accurate measures, performing a measurement system analysis (MSA) is quite important. MSA is a method for evaluating how much variation is present in the process by which we collect data. It answers the question regarding the accuracy and reliability of data as well as the process of data collection. MSA also helps in evaluating the effectiveness of the data collection plan.

In the context of robust quality, we also should measure the metrics' DQ by following the approach provided in Chapter 2. MSA and DQ checks are steps needed as part of the cross-examination of data to get meaningful conclusions. Calyampudi R. Rao, an internationally renowned statistician, provided a checklist (Rao, 1997) for the cross-examination of data. Some aspects of this checklist include:

- How are the data collected and recorded?
- Is the measurement system well defined?
- Are the data free from recording errors?
- Are the data from reliable sources and trustworthy?
- Are there any abnormalities or outliers associated with the data?
- Is the sample size adequate? Are all factors for data collection considered?
- Do you have the right kind of data/are the data suitable for the intended purpose?
- Are the data the right kind?
- Are the data suitable for the intended purpose?

If we carefully look at the checklist that Dr. Rao provided, it is clear that the emphasis is primarily given to the DQ and measurement systems that we use for data collection.

After performing MSA on the metrics, the next step is to identify the critical variables that are impacting the metrics. For these variables as well, MSA and DQ checks should be performed as needed.

The most important activity related to DQ is DQ assessment. This is an iterative effort involving acts of processing the results, validating them with the subject matter experts (SMEs), and refining the DQ rules that drive those results as needed. This work takes input from those who understand the datasets and the users of data.

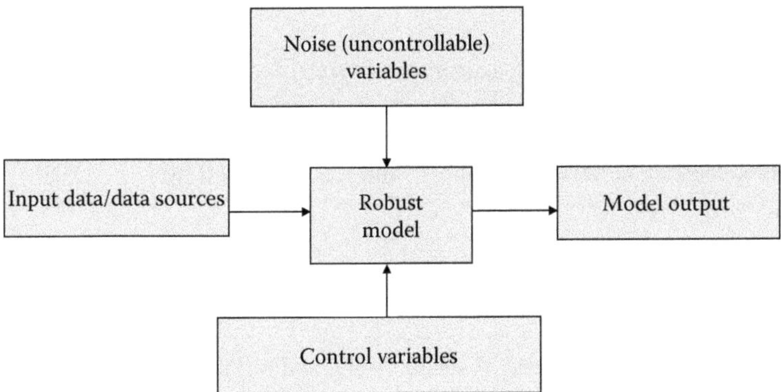

FIGURE 5.5 P-diagram representation.

Here, we need to answer questions such as "is a completeness score of 95% acceptable?" or "is 90% accuracy good enough?." Applying SPC concepts (Jugulum, 2014) is a good way to determine thresholds. These statistically determined thresholds should be adjusted as per subject matter experts (SMEs) inputs to obtain more practical thresholds to work with. Sometimes, when data are not available, we need to resort to SME inputs to begin with and then refine the threshold as we collect the data. After determining dimensional-level thresholds, a similar approach can be used for obtaining thresholds for overall DQ scores of CDEs. Readers can refer to Chapter 2 for the methodology to calculate overall DQ scores of CDEs. The next step in the Measure phase is to develop a conceptual analytical model with inputs and outputs. A parameter or p-diagram representation, as shown in Figure 5.5, is very useful here as well.

The p-diagram is very useful in clearly representing data sources, output from the model, and variables that are controllable and uncontrollable, which will help greatly when we have to fine-tune the variable settings for better output. Understanding which variables to control will also help in capability and potential analysis.

The Analyze Phase

After satisfying all of the data requirements of the Measure phase, we then have to decide on the type of analytics that we need to perform for the conceptual model. Depending on the need, a suitable type of analytical model/technique needs to be selected, as shown in Table 5.1. There are several techniques/tools that can be used in the development of analytical models. They include artificial intelligence (AI) techniques, genetic algorithms, machine learning and fuzzy logic techniques, artificial neural networks, decision trees, cluster analysis, factor analysis, regression analysis, principal component analysis, discrimination and classification analysis, time series analysis, correlations, association analysis, tests of additional information (Rao's test), and Taguchi methods. A brief description of some of these multivariate techniques is provided in the following. These are commonly used in analytical model development.

Correlation Analysis

Correlation analysis helps in measuring the degree of relationship between variables. To compute the linear correlation of a pair of variables, X and Y, we select two samples of X and Y, from which their correlation is calculated. If r_{XY} is the linear correlation coefficient between X and Y, and it is computed by:

$$r_{XY} = \frac{\text{Cov}\,(X,Y)}{s_X s_Y}$$

where:

$\bar{X}$ and $\bar{Y}$ are means of X and Y, respectively, and n is same size
s_X and s_Y are standard deviations of X and Y, respectively
cov (X, Y) is covariance between X and Y and is equal to $\sum (X_i - \bar{X})(Y_i - \bar{Y}) / (n-1)$
(i ranges from 1 to n)

The correlation coefficient lies between −1 and 1 with positive or negative values of 1 indicating perfect correlation. Depending on the values of the correlation coefficient, we can have strong positive or negative or mild positive or negative, or zero or no relationship between X and Y. Usually, the correlation information between the variables is represented in the form of a correlation matrix, as shown in the following table for the three variables X1, X2, and X3. Note that the correlation matrix is a symmetric matrix with diagonal elements equal to 1. In this example, the correlation coefficient between X1 and X2 is 0.7, that between X2 and X3 is 0.92, and so on.

Correlation Matrix (Example)

	X1	**X2**	**X3**
X1	1	0.7	0.8
X2	0.7	1	0.92
X3	0.8	0.92	1

Association Analysis

When the variables are of a discrete type, like processes, systems, regions, or types of employment, an association analysis is carried out on selected criteria to establish dependencies between the variables. Usually, a statistical test of significance is conducted to prove or disprove the hypothesis that there is a relationship or dependency between the variables. Using the test, a measured value of association is calculated and that value is compared with critical values to make decisions about dependencies.

Regression Analysis

When we want to establish the relationship between two variables (after ensuring that they are correlated) with mathematical equations, simple linear regression models are helpful. The regression equation is used to predict one variable (the dependent variable) based on the information available on another variable (the independent variable). A simple linear regression model would be of the following form:

$$Y_i = \beta_0 + \beta_1 X + \varepsilon$$

where:

β_0 and β_1 are regression constants evaluated by the method of least squares and ε is the model error.

From the historical data on X and Y, the regression model is constructed and that is used for predicting Y based on future observations of Xs.

If Y is dependent on k variables, $X_1, X_2, \ldots X_k$, then multiple regression analysis is used. A typical multiple regression equation would be of the form:

$$Y = \beta_0 + \beta_1 X_1 + \beta_2 X_2 + \ldots + \beta_k X_k + \varepsilon$$

The regression constants (βs) are estimated by the method of least squares and ε is the model error.

Stepwise Regression

By using stepwise regression analysis, we can identify the variables of importance required for the model. This is an iterative approach where in a sequence of regression models are constructed in steps by the addition and removal of variables based on some statistical criteria. There are many references for different types of regression analysis that readers can refer to.

Test of Additional Information (Rao's Test)

The test of additional information or Rao's test was developed by Calyampudi R. Rao to identify the important variables required for the model. In this method, a significance test is carried out for a subset of variables using Fisher's linear discriminant function. If the calculated value of the test statistic is not high, then that subset of variables can be discarded from the model. A high value of test statistic indicates that the variables used for the test provide additional information and are needed for the model.

Discrimination and Classification Method

The discrimination and classification method is intended for classifying objects or observations into different groups. A linear discriminant function is developed from historical data and is used for future classifications. This approach aims at the development of the discriminant function in such a way that groups of observations are separated as much as possible. For a detailed discussion on this method, please refer to Johnson and Wichern (1992).

Principal Component Analysis

Principal component analysis is typically used to reduce the dimensionality of the system by calculating linear combinations of original variables. The linear combinations, called principal components, explain the variance–covariance structure of the system. For a detailed discussion on principal component analysis, please refer to Johnson and Wichern (1992).

Artificial Neural Networks

Artificial neural networks consist of a diverse family of networks and are typically used in pattern recognition and classification analysis efforts. An artificial neural network is a structure with interconnected units. These units execute local computations and will have inputs and outputs.

Artificial Intelligence and Machine Learning Techniques

Artificial intelligence and machine learning techniques are intended to make a computer-related product think, understand issues, and provide solutions intelligently in a way similar to the manner that we humans perform these activities. These techniques are becoming increasingly popular in decision-making activities using analytics and many companies are making big investments in these areas. The idea of model performance risk management discussed at the end of this chapter can also help to evaluate the effectiveness of using analytical models including artificial intelligence and machine learning techniques and the investments made into them.

The Improve Phase

The Improve phase of the DMAIC approach focuses on validating the models with actual outcomes and making adjustments to them as needed. The outcomes should be practical and also should be analytically sound. This phase also focuses on the creation of visual displays in the form of dashboards and scorecards. At the end of this phase, we should have a clear plan for deploying insights, with appropriate steps.

The Control Phase

The Control phase should focus on monitoring and controlling model outputs and corresponding impacting variables. The use of SPC charts are highly recommended in the Control phase. In the Control phase, we should also periodically make sure that DQ levels are maintained, and we might need to re-run models on a periodic basis with new information and check for model validity. This is to make sure that high-quality data and insights are always maintained.

5.5 PURPOSEFUL ANALYTICS[1]

The analytics should have a purpose and their usage should be threefold: one aspect should focus on scientific discovery and advancement; the second aspect should focus on business decision-making through insights; and the third and most important aspect should focus on human welfare and the contribution to the society. Whatever aspect we deal with, the creation and use of insights through analytics should be purposeful. The use of the term *purposeful* in this regard is based on the following characteristics:

- Individualized insights
- Simplicity of usage
- Scalability
- Rapid insights
- Insights accuracy and precision
- Increased productivity

Many successful companies are investing millions of dollars each year into creating advanced analytics capabilities. These investments are aimed at developing both big data infrastructure with advanced techniques and highly skilled data and analytics experts. However, most often, these companies do not focus on the characteristics of *purposeful analytics*, although they are always of great interest to business leaders, who are always looking for ways to get quick results so as make judicious decisions in a cost-effective manner.

In this section, we discuss the importance and various aspects of purposeful analytics, including why this concept is so important in the data-driven world to achieve robust quality in analytics and in decision-making activities.

Figure 5.6 shows various characteristics of purposeful analytics. Their descriptions are provided in the following text.

Individualized insights: The insights derived through analytics should be individualized or personalized as necessary so that we can understand behavioral aspects at the individual level rather than at the group- or population level. Most analytical approaches cannot provide individualized level insights, as they are often based on population analytics. Individualized level analysis is very important so as to understand the drivers for all individuals. This will help us to consider suitable decisions based on individual requirements rather than on a group of similar individuals.

---

[1] The author is thankful to the Kanri, Inc. team, as he worked with them while developing some aspects of this section.

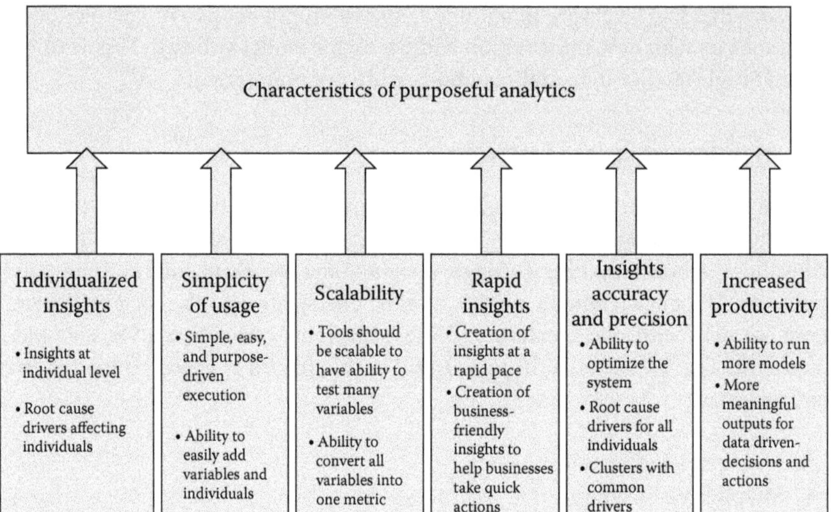

FIGURE 5.6 Characteristics of purposeful analytics.

Simplicity of usage: This refers to the importance of having simple and easy-to-run models as well as easy-to-interpret model outputs. This aspect is extremely important, as business decisions are made by business leaders who might not have exposure to advanced analytics or statistics. Simple and easy-to-run models must also have the ability to easily add variables and individuals.

Scalability: Scalability is measured by the ability to test many variables at the same time with different forms of data, numerical, text, voice, and so on, and by the ability to convert all variables into one metric. The tool that is being used should satisfy scalability requirements in performing analytical operations.

Rapid insights: Speed is a very important aspect of analytics, particularly if the organizations are making data-driven decisions. The creation of business-friendly insights at rapid pace is absolutely required to help businesses take quick actions. Obtaining rapid insights will also enable one to run more analytical models to generate more valuable information for making good decisions.

Insights accuracy and precision: It is extremely important to obtain insights precisely and accurately at the individual level by understanding key drivers affecting performance. In order to get key drivers, the analytical tool should have the ability to optimize the system by reducing the number of variables. Based on root cause drivers, one should be able to create clusters with common drivers so that suitable programs can be developed for the clusters.

Increased productivity: Once we have an innovative tool that is simple and easy-to use, we can enhance overall decision-making activity by running more models. This will help to generate more meaningful outputs for data-driven decisions and actions, which will translate into productivity gains by lowering the costs and resources.

Individualized Analytics versus Population-Based Analytics

Personalized/individualized insights are very important for purposeful analytics, as they are targeted for each individual separately rather than for all individuals put together in a population or group. The outputs of individualized analytics should focus on every individual rather than on a population of similar individuals, since the issues related to each individual are unique from one another due to different reasons. Individualized solutions will help organizations to take suitable actions that will enhance satisfaction at the individual level. If the number of individuals is high, then we can identify the variables or combinations of variables that are causing the problem or abnormal behavior and corrective actions can be planned accordingly. Table 5.2 summarizes some of the key differences between individualized analytics and population-based analytics.

Examples of Individualized Insights

If the objective is to identify variables that are impacting the nonperforming branches of a commercial bank, then Figure 5.7 shows the distances of 45 nonperforming branches from the desired state. From this figure, we can see that branches 10, 13, 29, 42, and 44 have higher distances and so should be looked into first.

Table 5.3 shows root cause drivers with percentage contributions to the distance for the top 10 branches. Note that contributions of 10% or greater are highlighted. In this table, for example, for branch 24, X9 drives only 6.7% and X5 is the key driver, with a 17.2% contribution. Similarly for branch 44, 30.5% of the distance is contributed by X4. This type of analysis will help us to understand and reduce the gap between actual performance and target performance of impacting variables. The reduction in this gap eventually reduces overall distance.

TABLE 5.2
Individualized Analytics versus Population-Based Analytics

Individual Analytics	Population-Based Analytics
Not a classification problem	Classification problem
Most often, the goal is to see how far individuals are from the target state	The same coefficients and equation are used for everyone
Root cause drivers for individuals separately	Root cause drivers for the group or population of individuals
Groups (clusters) based on variable contributions to the individuals. Specific improvement programs can be developed for these clusters	Start with groups (clusters) and put individuals in them and improvement programs are developed accordingly

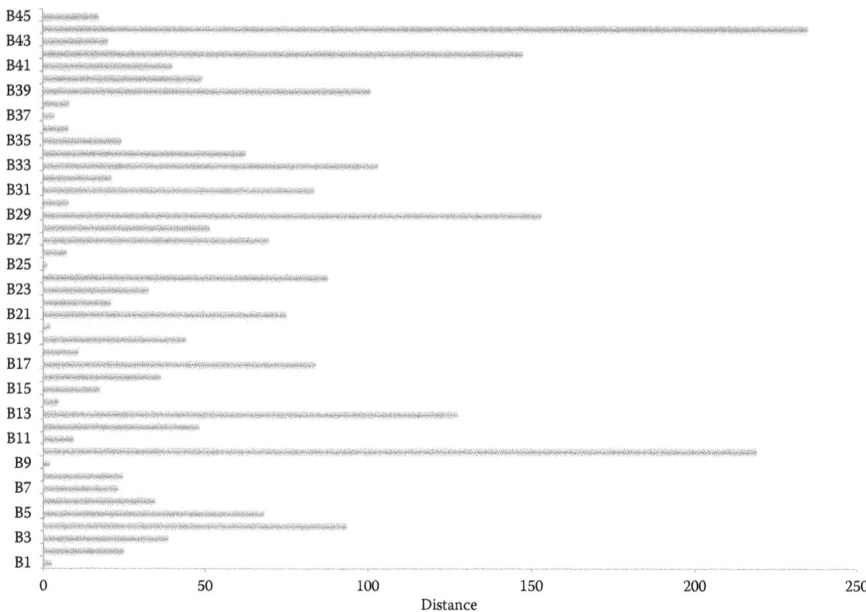

FIGURE 5.7 Branch distances from the desired state.

TABLE 5.3
Root Cause Analysis with Contribution Ratios

Branch #	Distance	X1	X2	X3	X4	X5	X6	X7	X8	X9	X10
B44	235.17	8.63	3.87	1.84	30.52	1.92	2.38	8.03	0.00	34.16	8.33
B10	219.26	8.33	5.05	6.31	17.62	7.20	6.56	9.40	13.30	8.13	12.14
B29	153.24	7.68	1.35	5.22	10.08	2.26	8.42	2.69	2.02	49.09	9.39
B42	147.54	12.58	5.30	6.81	16.98	21.25	5.03	4.25	3.36	7.35	11.71
B13	127.71	10.84	1.64	3.18	11.48	5.99	0.79	2.59	20.43	29.78	5.78
B33	103.43	4.75	0.27	0.48	9.24	7.07	5.52	3.78	4.79	49.19	9.43
B39	101.09	20.49	0.01	12.98	26.54	1.18	0.70	0.37	1.72	23.51	10.00
B4	93.69	4.49	2.36	5.41	5.36	0.07	3.46	14.56	1.85	52.38	7.17
B24	87.82	8.20	6.64	5.58	10.59	17.27	9.03	7.98	10.49	6.69	11.80
B17	84.10	2.44	0.00	0.17	22.16	2.52	0.17	2.68	0.47	63.09	5.69

5.6 ACCELERATED SIX SIGMA FOR PROBLEM-SOLVING

Six Sigma process engineering effort can be accelerated by using a suitable multivariate tool, as it helps in collectively testing the effects of several variables at once. By selecting a suitable tool and approach, overall effort can be minimized to a great extent by arriving at important drivers very quickly, and this could result in huge reductions in time and costs. The steps required for accelerated

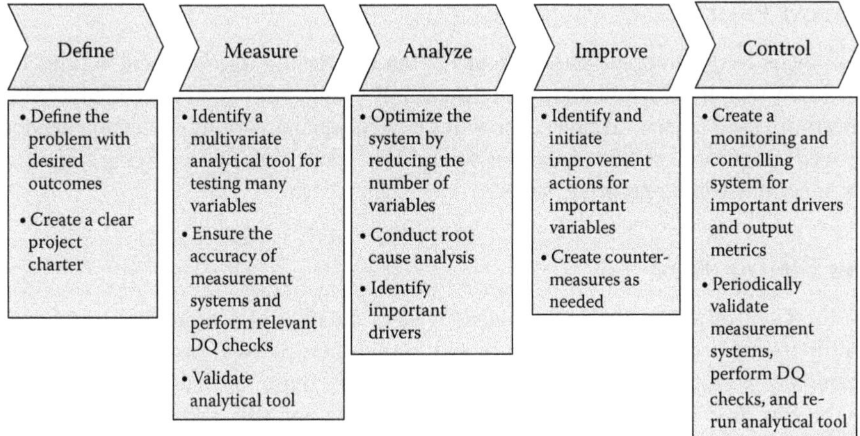

FIGURE 5.8 DMAIC phases for accelerated Six Sigma.

Six Sigma efforts can fall into various phases of the DMAIC approach, as shown in Figure 5.8. The descriptions of the DMAIC phases for accelerated Six Sigma are given henceforth.

The Define Phase

In this phase, the problem must be clearly defined, along with the desired outcomes. The project team must clarify the roles related to the project and identify the individuals for these roles. The creation of the project charter with a scope, objectives, resources, expected business value, and role clarity is vital in this phase. A detailed project plan for the five phases of DMAIC with the tasks and associated deliverables included is also important in this phase.

The Measure Phase

The focus in the Measure phase should be on the selection of a suitable multivariate tool that can test several hypotheses with reference to the target performance. After selecting a suitable tool, the focus should be on data collection for the variables selected. We should also ensure the accuracy of measurement systems and perform relevant DQ checks to ensure high DQ. After this step, the tool should be validated with known outcomes so that a high degree of confidence can be established regarding the selected tool.

The Analyze Phase

After validating the analytical tool, the multivariable system should be optimized by identifying key variables. Root cause analysis must be performed on these key variables to identify important drivers of the problem.

Improve Phase

The focus in the Improve phase should be on identifying improvement actions for the important drivers. Suitable countermeasures should be created as needed. If SPC charts are used for the variables, one will know the quantum of improvement needed as well. At the end of this phase, we should have clear plan for deploying improvement actions with appropriate steps.

The Control Phase

In the Control phase, the important activity is creation of a monitoring and controlling system for important drivers and output metrics. SPC charts are typically extremely useful in the Control phase. During this phase, one should periodically make sure that measurement systems are accurate and that DQ levels are maintained, in addition to ensuring the re-running of the analytical tool. This is required in order to make sure that high-quality data and insights are always maintained.

5.7 MEASURING ANALYTICS QUALITY

In this section, we will discuss how we can measure the robust quality of analytics execution by using the analytics robust quality index (RQI), which takes into account both analytics quality and DQ aspects. The procedure for calculating RQI is presented in Chapter 4. Note that here also, RQI lies between 0 to 100 with 100 representing the best model. Consider three CDEs/factors as required inputs for an analytical model. For these three factors, we need to examine DQ levels by defining suitable DQ dimensions. Table 5.4 shows an illustrative example with DQ dimensions, along with corresponding scores and overall DQ scores.

After obtaining DQ scores for the metrics, in the next step, we need to examine the analytics quality from the model outputs. For measuring RQI for analytics execution, it is required to calculate deviations of DQ scores from the target and analytics outputs from the actual values. Since DQ scores are measured on a scale of 0 to 100, it would be better to transform analytical outputs to the scale of 0 to 100. After doing this and defining targets, deviations can be calculated, and, from the deviations, signal-to-noise ratios are calculated. Based on the signal-to-noise ratios, analytics RQIs are measured on a scale of 0 to 100, with a score of 100 indicating the best model. Table 5.5 shows details of analytics RQI calculations for one model. The RQI for this model is 44.3 which is in the middle range.

TABLE 5.4
DQ Evaluation of Three CDEs/Factors

DQ Dimension	Completeness	Conformity	Validity	Accuracy	DQ Score
CDE1/Factor 1	96%	93%	88%	87%	91%
CDE2/Factor 2	95%	90%	85%	78%	87%
CDE3/Factor 3	92%	88%	84%	76%	85%

TABLE 5.5
Analytics RQI for a Model

CDE/Factor	Score	Target	Squared Deviation
CDE1/Factor 1-DQ	91	95	16
CDE2/Factor 2-DQ	87	90	9
CDE3/Factor 3-DQ	85	90	25
Model output	75	100	625
		MSD	13.0
		SNR	−11.1
		RQI for analytical model	44.3

MSD: Mean square deviation; SNR: S/N ratio; CDE: Critical data element; DQ: Data quality; and RQI: Robust quality index.

TABLE 5.6
Overall Analytics RQI

	CDE	Score	Target	Squared Deviation
Model 1	CDE1/Factor 1-DQ	91	95	16
	CDE2/Factor 2-DQ	87	90	9
	CDE3/Factor 3-DQ	85	90	25
	Model output	75	100	625
Model 2	CDE1/Factor 1-DQ	86	90	16
	CDE2/Factor 2-DQ	93	96	9
	Model output	70	100	900
			MSD	15.12
			SNR (log)	−11.80
			RQI for analytics	41.02

MSD: Mean square deviation; SNR: S/N ratio; CDE: Critical data element; DQ: Data quality; and RQI: Robust quality index.

If there are multiple models, then overall analytics quality can be obtained by using RQI results of all models. Table 5.6 illustrates the method of obtaining overall analytics quality in a two-model scenario. The overall RQI for an analytics function with these two models is 41.02.

5.8 MODEL PERFORMANCE RISK MANAGEMENT USING ANALYTICS ROBUST QUALITY INDEX

In this analytics-driven world, it is very important to have a systematic, rigorous, and scientific way of assessing the performance of models to ensure that we are generating valuable insights that drive several organizational decisions. There is a big gap

TABLE 5.7
RQI for Model Performance Evaluation

Criterion	Score	Target	Squared Deviation
Precision	90	100	100
Value	75	90	225
Ease of use	80	90	100
Actionable	70	80	100
Comparable	85	100	225
		MSD	12.3
		SNR	−10.9
		RQI for model performance	45.6

RQI: Robust quality index; MSD: Mean square deviation; and SNR: S/N ratio.

in this area in many analytically driven companies. When there is formal criteria to evaluate model performance, analytics ROI is very useful in assessing the model performance.

Let us suppose that the models are assessed based on scores of the following criteria:

Accuracy: Insights quality (consistency) through measurement system analysis
Value: Impact of model through cost benefit analysis
Ease of usage: Simplicity of model
Actionable: Ability to implement
Comparable (Accuracy): Closeness to actual outcomes

If we know the values of these aspects along with target scores, then we can calculate the analytics RQI to assess the model's performance. Table 5.7 shows an example of this. From this analysis, it is clear that the model performance score is 45.6, which is in the middle range.

It is important for organizations to develop model performance risk management framework by using evaluation indices like RQI, and such a framework should be integrated into the organization's overall risk management policy. In addition, the model risk management framework should periodically be updated and evaluated by involving an independent team of experts. Model performance risk management will also help to evaluate the effectiveness of using analytical models including artificial intelligence and machine learning techniques and the investments made on them.

6 Case Studies

This chapter presents several case studies from different industries. These case studies show the importance of addressing data quality (DQ) and process quality aspects in achieving the overall quality of products and services.

6.1 IMPROVING DRILLING OPERATION[1]

A printed circuit board (PCB) is a base component with electrical interconnections on which several components are mounted to give the desired electrical output. Drilling is one of the most important operations of PCB manufacturing. Drilled holes are considered very important, as connectivity is made through them. Drilled holes produce an opening through the board to permit subsequent processes to form electrical connections as well as enable component mounting with structural integrity and precision of location. Typically, drilling operations are carried out on stacks of copper panels that consist of PCBs of specific sizes and shapes with one or more circuits. Copper panels have substrate layers in between copper layers. The number of panels in a stack that is used in drilling is referred to as the stack height. A stack may contain two, three, or four panels. There were several quality issues with drilling identified and therefore a project was undertaken.

Drilling Defects

Drilling defects can be classified as copper defects and substrate or epoxy defects. The defects under each category are as shown in Table 6.1. The definitions (Coombs, 1988) of these defects are presented in Tables 6.2 and 6.3.

Measurement System Analysis

As part of the Measure phase activities, a measurement system was sourced through an extensive literature search. Finally, a suitable methodology was identified and selected from (Coombs, 1988). This measurement selection approach was essential, since quality improvement could not be carried out without a suitable quantitative measurement method. According to this method, the bottom panel is always used for the analysis of hole quality. Holes are designed on the panel in such a way that four holes are drilled after drilling a certain number of holes (also referred to as the number of hits). One hit was equivalent to drilling one hole. The holes used to measure DQ are known as *coupon holes*. This is demonstrated visually in Figure 6.1.

[1] The author would like to thank R. C. Sarangi and A. K. Choudhury for their help and support. Thanks as well are due to Wiley for granting permission.

TABLE 6.1
Drilling Defects

Copper Defects	Substrate Defects
Delamination	Delamination
Nail heading	Voids
Smear	Smear
Burr	Plowing (roughness)
Debris	Debris pack
Roughness	Loose fibers

TABLE 6.2
Measuring Hole Quality (Copper Defects)

Defects	Definition	Weightage Factor (a_i)	Extent Factor (b_i)
Burr	Burr height (microns)		
(a ridge on the outside	1.524	1.0	0.01
of the copper surface	4.864		0.08
after drilling)	7.874		0.30
	12.954		1.20
Nail heading	Nail head width (microns)		
(a flared condition of	3.040	1.5	0.01
the internal	5.440		0.08
conductor)	15.740		0.30
	25.900		1.20
Smear	Percentage of copper area		
(fused deposit on the	covered with smear		
copper from	1%	1.5	0.01
excessive drilling	11%		0.08
heat)	26%		0.30
	36%		1.20

The coupon holes are removed from the bottom laminate and are then molded by thermosetting plastic material. The holes in the mold undergo grinding and polishing activities until the holes are opened to the necessary diameter. After these steps, the holes are measured by using a microscope. The coupon holes are usually seen through the microscope under different magnifications, as defined in Tables 6.2 and 6.3.

This methodology satisfies the requirements for performing measurement system analysis (MSA). As we know, MSA helps with evaluating how much variation is present in the process by which we collect data. It addresses the accuracy and precision of the data. In this case, it was done via repeated measurements of hole quality, which was determined as follows:

$$\text{Hole quality} = 10(0.2)^{\Sigma ai\ bi}$$

TABLE 6.3
Measuring Hole Quality (Substrate Defects)

Defects	Definition	Weightage Factor (a_i)	Extent Factor (b_i)
Void (a cavity in the substrate)	Minimum magnification required to see the defects clearly		
	140X	0.8	0.01
	100X		0.08
	60X		0.30
	20X		1.20
Debris pack (debris deposited in cavities)	Percentage substrate area covered with debris		
	1%	0.8	0.01
	11%		0.08
	26%		0.30
	36%		1.20
Loose fibers (supporting fibers in the substrate of a laminate that are not held in place by the surrounding resin)	Percentage substrate area covered with loose fibers		
	1%	0.3	0.01
	11%		0.08
	26%		0.30
	36%		1.20
Smear (fused deposit left on the substrate from excessive drilling heat)	Percentage of the substrate area covered with smear		
	1%	0.3	0.01
	11%		0.08
	26%		0.30
	36%		1.20
Plowing (furrows in the hole wall due to drilling)	Minimum magnification required to see the defect clearly		
	140X	0.2	0.01
	100X		0.08
	60X		0.30
	20X		1.20

X: Times magnification using a microscope.

where $\Sigma a_i b_i$ equals the sum of the product of weightage factors and the extent factors of all defects.

If the extent factors are not given in Tables 6.2 and 6.3, they can be determined by interpolation.

Hole quality lies between 0 and 10. Any value above 6.0 is considered satisfactory.

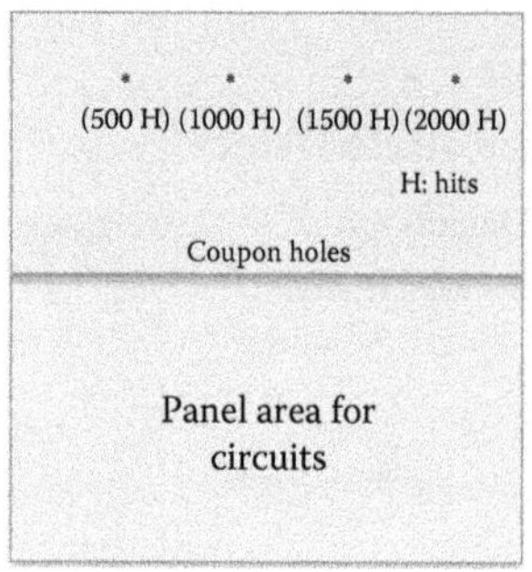

FIGURE 6.1 Coupon holes on a panel.

TABLE 6.4
Existing Hole Quality

Copper Defects	Value	a_i	b_i	$a_i\ b_i$
Burr height	17.5 microns	1.0	2.0	2.0
Nail head width	10.0 microns	1.5	0.245	0.367
Smear	21.42%	1.5	0.267	0.400
Substrate Defects				
Voids	100X	0.8	0.08	0.064
Plowing	100X	0.2	0.08	0.016
Smear	5.46%	0.3	0.04	0.012
Loose fibers	0.468%	0.3	0.005	0.0015
Total				**2.8605**

X: Times magnification using a microscope.

The first task in the MSA phase is to estimate the existing hole quality. The details of the existing hole quality measurements along with weightage factor and extent factors are given in Table 6.4.

$$\text{Hole quality} = 10\left(0.2^{2.8605}\right) = 0.100$$

The hole quality image as measured (through a microscope) is shown in Figure 6.2. As can be seen, the existing hole quality was very low, with irregular rough hole walls and many defects.

Since the quality of drilling was so poor, the process needed complete reengineering, so the choice was made to study impacts of various factors on drilling by using experimental design.

Experiment Design Description

As we know, for designing experiments, it is important to understand all possible factors and their impact on the output. The overall drilling process system with all relevant factors is as shown in a p-diagram in Figure 6.3.

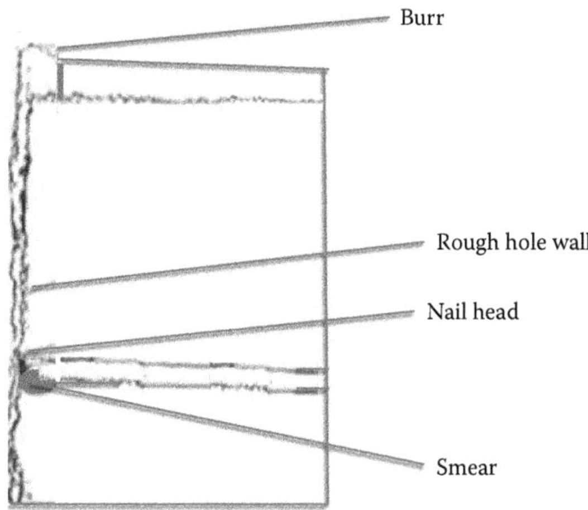

FIGURE 6.2 Image of existing hole quality.

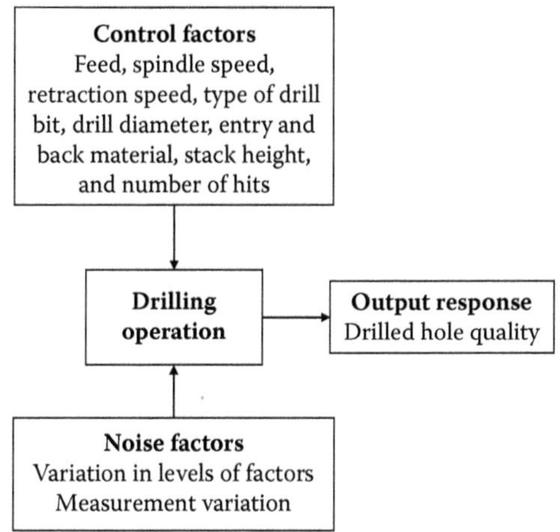

FIGURE 6.3 P-diagram for drilling operation.

Selection of Levels for the Factors

For a given drill, spindle speed and retraction speed are usually fixed. Since four layer boards with 0.95 mm diameter drill bits were considered for the purpose of this experiment. The factors and levels shown in Table 6.5 were also considered for the purpose of this experiment.

In Table 6.5, Factor B was the supporting material for the panels to facilitate the drilling operation. Factor C is the number of panels drilled at one time. Factor D is the number of holes drilled per drill bit. Factor E corresponds to the type of drill bit. Ordinary drill bits and neck-relieved drill bits were considered in this study. Neck-relieved drill bits, also known as undercut drill bits (see Figure 6.4),

TABLE 6.5
Factors and Levels

Factor	Level 1	Level 2
Feed (A)	121 IPM	138 IPM
Entry and back-up material (B)	Aluminum	Phenolic
Stack height (C)	3 high	4 high
Number of hits (D)	2,000	1,500
Drill bit type (E)	Ordinary (OD)	Neck-relieved (NR)

IPM: Inches per Minute.

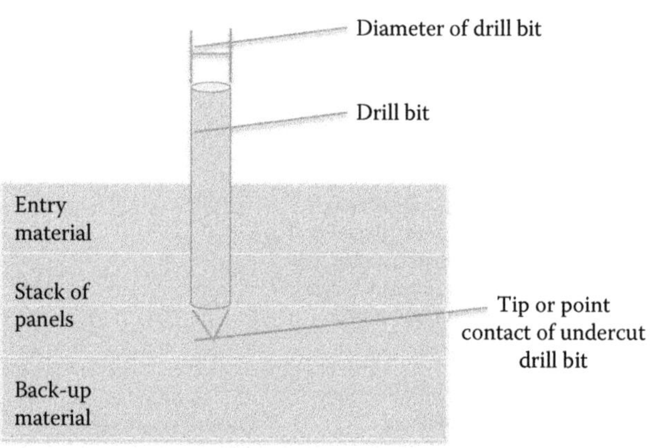

FIGURE 6.4 Undercut or neck-relieved drill bit.

have a different shape at the tip so that the only point of contact is where the drill bit touches the panel. Because of this, all of the heat generated during drilling will not get transferred to the hole wall all at once, resulting instead in a gradual distribution of heat.

Since a drilling operation is carried out in a controlled environment, external noise factors were not considered in this experiment. The noise factors that were considered were noise due to change levels and noise due to measurement variation.

Designing the Experiment

An orthogonal array with eight experiments ($L_8(2^7)$ array) was chosen to accommodate five factors for the experiment. This array was adequate, since required effects (all main effects and interaction effects) can be estimated with it. The factors were allocated to the columns of the $L_8(2^7)$ array. The layout of the experiment is as shown in Table 6.6.

TABLE 6.6
Physical Layout of the Experiment

Test	A	B	C	D	E
1	121 IPM	LCOA	3 High	2,000 hits	OD
2	121 IPM	HYLAM	4 High	2,000 hits	NR
3	138 IPM	Aluminum	4 High	2,000 hits	OD
4	138 IPM	Phenolic	3 High	2,000 hits	NR
5	121 IPM	LCOA	4 High	1,500 hits	NR
6	121 IPM	HYLAM	3 High	1,500 hits	OD
7	138 IPM	LCOA	3 High	1,500 hits	NR
8	138 IPM	HYLAM	4 High	1,500 hits	OD

IPM: Inches per Minute.

Data Collection and Ensuring Data Quality

For each experiment, hole quality was measured four times. For all experiments, DQ tests were performed to ensure the validity of the data (hole quality should lie between 0 and 10) and to conform to the standards based on the selected measurement system. Since this was an experimental design and four observations were collected for each test, the completeness of the data was also ensured (though this might not be so relevant in this case since the sample size is small and can be easily measured). The accuracy was measured based on the output from the microscope. Thus, four important DQ dimensions—completeness, validity, conformity, and accuracy—of the data were addressed before analyzing the experimental results. The target value for accuracy was considered to be 97% and the actual accuracy value was considered to be 94%. All other dimensional scores were 100%

Data Analysis

Since our goal was to maximize the hole quality, the difference between the observed hole quality and the target value (10) was considered as the response. Accordingly, smaller-the-better-type signal-to-noise (S/N) ratios were considered for the analysis. The results of experiment and corresponding S/N ratios are shown in Table 6.7.

The results of the experiment were analyzed based on S/N ratios. The details of the analysis are shown in Table 6.8. The average response corresponding to these effects can be seen from Figure 6.5.

Through the statistical analysis, the optimal combination was selected and is as follows:

Feed (A2)	138 IPM
Entry and back-up (B1)	Aluminum
Stack height (C1)	3 high
Number of hits (D2)	1500
Drill bit type (E2)	Neck-relieved

TABLE 6.7
Results of the Experiment

Test	Values of Hole Quality	S/N Ratio (dB units)
1	0.22, 0.146, 1.498, 1.348	–19.293
2	4.007, 4.008, 4.318, 4.678	–15.119
3	0.03, 0.100, 0.17, 0.10	–19.964
4	4.383, 4.945, 4.652, 5.11	–14.378
5	4.01, 4.01, 4.01, 4.02	–15.544
6	0.318, 0.267, 2.23, 2.29	–18.868
7	6.4, 6.0, 6.5, 6.3	–11.375
8	0.152, 0.318, 1.21, 1.35	–19.330

S/N: Signal-to-noise.

TABLE 6.8
Analysis with S/N Ratio as the Response

Source	Degrees of Freedom	Sum of Squares	Mean Squares	Contribution Ratio (%)
A	1	1.859	1.859	0.684
B	1	1.619	1.619	0.322
C	1	4.686	4.686	4.95
D	1	1.727	1.727	0.485
E	1	54.909	54.909(*)	80.687
Error	2	2.81	1.405	12.872
Total	7	67.61		

(*)Highly significant.

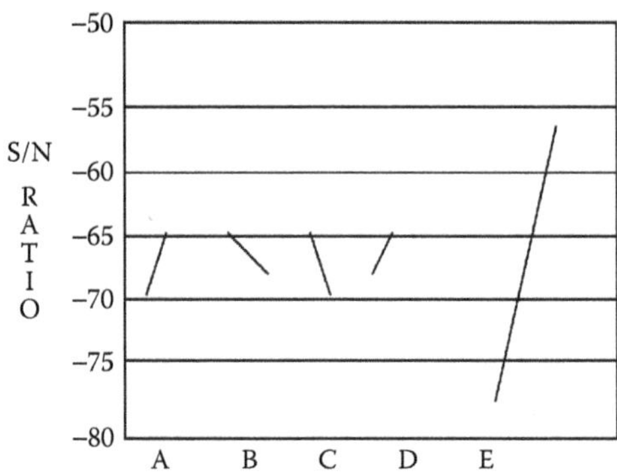

FIGURE 6.5 Average responses.

TABLE 6.9
Hole Quality - RQI Improvement

Before	Scores	Target	Squared Deviation
Hole quality (DQ)	94	97	9
Hole quality (PQ)	1	100	9801
		MSD	70.03
		SNR (log)	−18.45
		RQI	7.73
After	**Scores**	**Target**	**Squared Deviation**
Hole quality (DQ)	94	97	9
Hole quality (PQ)	64	100	1296
		MSD	25.54
		SNR	−14.07
		RQI	29.63

DQ: Data quality; PQ: Process quality; RQI: Robust quality index; MSD: Mean square deviation; and SNR: S/N ratio.

Confirmation Experiment

Since optimal combination was not part of the orthogonal array experimental layout, a confirmation experiment was performed for the optimal combination, which provided an S/N ratio of −52.39 dB and a corresponding hole quality of 6.4.

Improvement in Robust Quality Index

Since accuracy of measurements was considered at 94% against 97% target, the DQ component of robust quality index (RQI) was measured with this information. The hole quality before the experiment was 0.1 on a scale of 0 to 10. For this scenario, RQI was 9.81. RQI after the improvement in hole quality (which was 6.4) with the same accuracy of measurements was 29.63. This was considered huge in comparison with the prior situation, although there was still room for more improvement. Table 6.9 shows the details of RQI improvement.

The optimal combination was implemented in actual production, as the results were satisfactory.

6.2 IMPROVING PLATING OPERATION[2]

This study (Chowdhury, A.R., Jugulum, R., and Prasad, T. (1998)) was also carried out in a PCB manufacturing industry and was related to electroplating operation. In the electroplating process, the circuitry is plated with copper for electric connectivity.

[2] The author would like to thank R. C. Sarangi, A. K. Choudhury, G. Krishna Prasad, and A. R. Chowdhury for their help and support. Thanks are also due to Springer for granting permission.

For this process, the copper deposition rates were observed to be varying to a great extent, which was not desirable. The adverse effects of improper (nonuniform) plating were poor solderability, low ductility, and low mechanical strength. Reducing these defects was essential for the repeated component replacements and long life of PCBs. Therefore, it was important to keep variation due to plating thickness as low as possible. The specification width of thickness inside the hole was 25–35 microns. The existing electroplating process performance was studied and the process performance was deemed to be unsatisfactory with high thickness levels with a mean of 32 microns and a high standard deviation of 8 microns.

Validating the Measurement System

To identify and examine the nature and sources of overall plating thickness variation, a nested design was used. The data were collected through a hierarchical design, as shown in Figure 6.6. The plating operation was carried out in any of the 10 tanks numbered from 1 to 10, which were classified into four stages, as shown in Figure 6.6. In this figure, stage I and stage II had three tanks, while stage III and stage IV had two tanks. From each tank, two panels were considered and the thickness was measured at the following three positions: top, center, and bottom. A panel consists of one or more circuits, which would be routed to the required shape during the routing operation. In each position, two observations were recorded. After collecting the data in this format, an analysis was performed. From the analysis, it was found that the variation due to the different stages and the variation due to the different tanks were not significant. However, variation from panel to panel and variation between positions within the panels were significant. This type of measurement system validation is extremely important to understand overall variability, significant variation sources, and where measurements need to be collected for improvement activities. Table 6.10 provides the details of the analysis.

As the primary objective was to minimize the variation from panel to panel and the variation within the panels, various aspects of the plating operation were considered. A parameter diagram or p-diagram for plating operation shows these aspects (Figure 6.7).

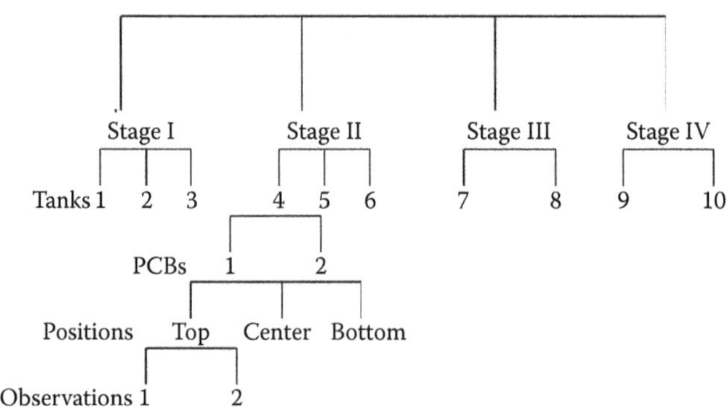

FIGURE 6.6 Hierarchical nested design.

TABLE 6.10
Analysis of Nested Design

Source	Degrees of Freedom	Sum of Squares	Mean Squares
Stage	3	351.48	117.16
Tank	6	246.27	41.04
Panel to panel	10	911.33	91.13(*)
Within panel	40	176.56	4.41(*)
Error	60	137.34	2.29
Total	119	1,822.98	

(*)Significant at 5% level of significance.

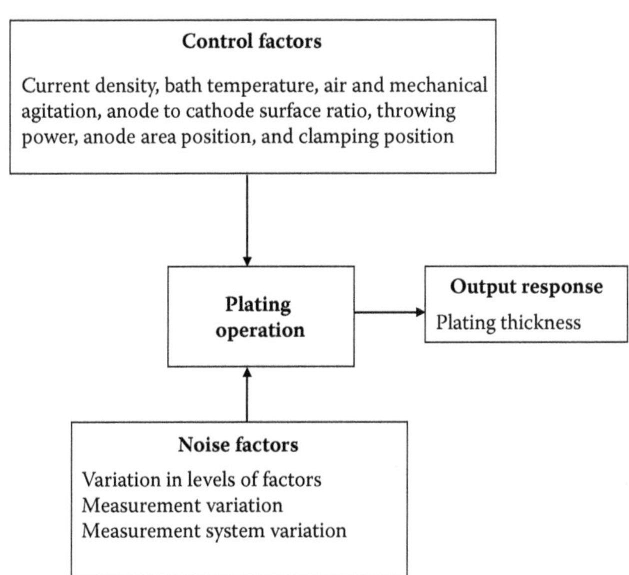

FIGURE 6.7 P-diagram for plating operation.

Out of these factors, two factors—clamping position and the number of anode bags—were considered to be critical. Therefore, an experiment was performed involving these two factors in order to find out the combination that gives thickness values to close to 30 microns with the least amount of variation. A description of these factors is given henceforth.

Anode Area and Position

As the current is passed to the anode, the copper pellets are dissolved in the electrolytic copper sulfate solution and deposited over the cathode (the panels). The copper pellets are stored in bags. To produce uniform copper plating, the distance between the anode bags to the panels is quite important. The distance and copper deposition rates were inversely proportional to each other, meaning the higher the distance, the lower the copper deposition and, the lower the distance, the higher the deposition. Therefore, for low plating variation, uniform distance must be maintained between the panels and the anode bags. With the existing arrangement containing four anode bags, it was not possible. Therefore, it was necessary to increase the number of anode bags to have uniform plating thickness over the panels. Based on the configuration of the plating tanks, it was decided to increase the number of anode bags to six and seven. Therefore, the levels for this factor were four bags, six bags, and seven bags.

Clamping Position

Suitable clamping position was required to reduce within-panel variation. The panel area closer to the clamps would typically form a low-current density and the area away from the clamps would typically form a high-current density. As a result, the plating would be thicker at the higher-density area and thinner at the lower-density area. This variation in the plating thickness could be reduced by properly positioning the clamps. So, the three clamping positions (levels) of single clamping, double clamping, and side clamping were selected for experimentation.

The factors and levels chosen for experimentation are shown in Table 6.11.

Designing the Experiment

With two factors each at three levels, it was decided to study all possible experimental combinations. The number of experiments required for this full factorial experiment was $3^2 = 9$. The physical layout of the experiment is shown in Table 6.12.

TABLE 6.11
Factors and Levels

Factor	Level 1	Level 2	Level 3
Anode bags (A)	4 bags	6 bags	7 bags
Clamping position (B)	Side	Single	Double

TABLE 6.12
Physical Layout of the Experiment

Experiment Factor	1	2	3	4	5	6	7	8	9
A	4 bags	4 bags	4 bags	6 bags	6 bags	6 bags	7 bags	7 bags	7 bags
B	Side	Single	Double	Side	Single	Double	Side	Single	Double

Data Collection and Data Quality

A full factorial experiment was conducted as per the plan in Table 6.12. For each combination, measurements of the panels were taken. Because of the configuration of the clamping position, the measurements were taken on six panels for side clamping and single clamping and on three panels for double clamping, respectively.

For each panel, thickness was measured at 10 positions in each of the panels. For all of the measurements, DQ tests performed to ensure the validity of the data and to conform to the standard based on the selected measurement system. Since this was an experimental design and 10 observations were collected for each test, the completeness of the data was also important. In this study, also, four important DQ dimensions—completeness, validity, conformity and accuracy—were addressed and DQ was ensured before analyzing the experimental results. The accuracy of the instrument measuring the thickness was considered to be 98%, although the target value was considered to be 100%. All other dimensional scores were 100%.

Data Analysis

The mean thickness and the standard deviation (SD) for each replication were calculated. The details are given in Table 6.13. The results were analyzed through analysis of variance and are summarized in Table 6.14.

The response graphs are shown in Figure 6.8.

TABLE 6.13
Mean and SD (in microns)

	Side Clamping		Single Clamping		Double Clamping	
	Mean	**SD**	**Mean**	**SD**	**Mean**	**SD**
4 anode bags	32.67	4.82	28.19	3.14	29.73	3.88
	38.30	7.80	21.44	1.57	31.67	2.95
	31 70	5.72	26.75	1.60	26.50	4.63
	31.41	8.63	21.26	1.18		
	26.61	7.58	18.24	2.92		
	33.51	7.91	26.16	2.80		
6 anode bags	28.20	3.74	25.95	2.36	29.29	2.92
	26.64	6.10	26.46	2.98	25.71	2.27
	32.90	4.09	29.65	2.68	24.20	2.68
	36.72	5.63	29.82	2.92		
	28.52	5.30	26.48	2.73		
	33.93	7.62	25.53	2.41		
7 anode bags	25.47	5.16	18.86	2.38	25.92	2.09
	26.54	4.17	24.02	1.62	30.93	2.08
	24.42	1.76	27.58	1.81	31.38	2.78
	30.21	3.94	29.77	2.33		
	28.76	7.03	30.86	3.48		
	30.89	7.74	25.20	2.32		

TABLE 6.14
Analysis of Variance

Source	Degrees of Freedom	Sum of Squares	Mean Squares
A	2	0.48	0.24(*)
B	2	5.32	2.66(*)
AB	4	0.57	0.142
Error	36	2.59	0.072
Total	44	8.82	

(*)Significant at 5% level of significance.

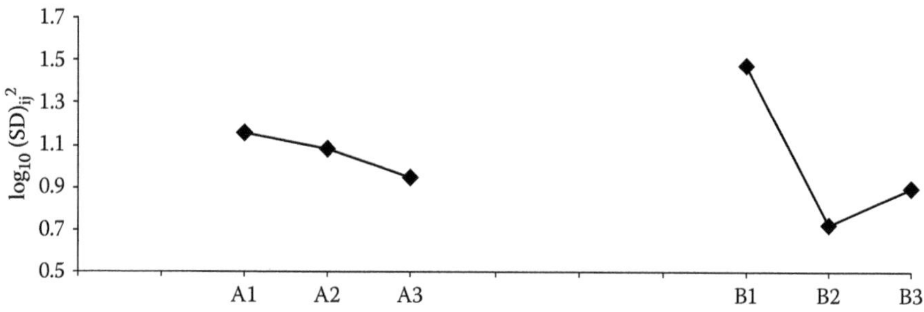

FIGURE 6.8 Response graphs.

From analysis of variance tables and response graphs, it was concluded that the effect of clamping position and the effect of the number of anode bags were significant. Also, from Table 6.13, we can see that seven anode bags and double clamping was the best combination, since the thickness values were closer to the target of 30 microns with lower standard deviation. Hence, seven anode bags and double clamping was selected as the optimal combination.

Calculating Robust Quality Index

Since the accuracy of the measurements was considered to be 98% against the target of 100%, the DQ component of RQI was measured with this information. The plating thickness for optimal combination (seven anode bags and double clamping) was 29.41 microns and the target was 30 microns. For this situation, RQI was calculated as shown in Table 6.15 and the value was 85.12. This value was much higher than that in the prior situation. The optimal combination was implemented into actual production by changing the method of loading the panels.

TABLE 6.15
RQI for Plating

	Scores	Target	Squared Deviation
Plating thickness (DQ)	98	100	9
Plating thickness (PQ)	29.41 microns	30 microns	3.87
		MSD	1.98
		SNR	–2.97
		RQI	85.12

DQ: Data quality; PQ: Process quality; RQI: Robust quality index; MSD: Mean square deviation; and SNR: S/N ratio.

6.3 DATA QUALITY IMPROVEMENT PRACTICE TO ACHIEVE ROBUST QUALITY[3]

A large bank established an enterprise level-data management function with the focus on DQ improvement and consistency across all business areas. This function was intended to drive data culture through proper data governance and systematic approaches so that a high quality of data and information is ensured to support all decision-making activities. This function also incorporated DQ measurement in its continuous efforts to identify, assess, and manage all types of risk (regulatory, market, operational, and so on). Sound DQ capabilities are necessary to increase customer and regulator confidence in the data that support the business. In addition, these capabilities enable the organizations to increase the effectiveness and efficiency of all operations.

As we know, the DQ assessment and improvement begins with identification of critical data elements (CDEs) to be monitored, controlled, and improved. To identify CDEs, a funnel methodology was developed so that the right CDEs are selected from a large pool of data elements. The funnel approach is described in Chapter 2. This particular study is related to the Basel II Accords/Standards. Many financial companies have begun to adopt the regulatory measures outlined in the Basel II Accords. These accords are intended to ensure that banks follow a disciplined process to set aside sufficient cash reserves in order to offset different types of risks incurred through their operational practices.

Critical Data Element Rationalization Matrix

In the beginning of this effort, the risk function of this bank started with a list of 35 data elements, from which CDEs would be selected. A CDE rationalization

[3] The author worked with C. Shi, H. I. Joyce, J. Singh, B. Granese, R. Ramachandran, D. Gray, C. H. Heien, and J. R. Talburt on this effort and sincerely thanks them. Thanks are also due to Wiley for granting permission.

Weightages / Criteria / CDEs	10	7	7	7	10	10	10	7	7	4		
	Ease of access	% of reports	% of customers	Total incidences	Business support	Technolgy support	Regulatory risk	Financial risk	Reputation risk	Operational risk	Total score	
CDE1	7	10	7	10	1	7	4	7	4	7	484	
CDE2	1	10	7	7	1	4	7	10	4	10	436	
CDE3	4	7	1	10	10	4	4	1	1	7	388	
CDE4	10	7	1	1	1	4	7	7	10	4	418	
.....	...	...	...	...	...	...	...	...	...	...	...	
CDE22	4	4	4	10	10	10	7	10	7	10	595	Highest
CDE23	4	4	1	4	1	10	4	4	7	1	292	
....	...	...	...	...	...	...	...	...	...	...	...	
CDE35	4	10	7	7	1	7	1	4	1	7	361	

FIGURE 6.9 CDE rationalization analysis matrix.

analysis was used for prioritizing data elements and selecting CDEs. Prioritization was done through scoring the CDE dependencies against weighted ranking criteria and deriving a total, which is a sum product of all criteria weights and their corresponding scores. This tool helped in conducting a high-level potential CDE-filtering that was based on science and experience. The range of total scores (223–595) provided a good means to decide on the criticality of data elements. Through this approach, the number of CDE candidates under consideration was reduced from 35 to 21. Figure 6.9 shows a sample CDE rationalization and prioritization analysis matrix.

Correlation and Association Analysis

In the next step of prioritization, statistical correlation and association analysis techniques were carried out on the information related to 21 remaining candidate CDEs. In cases in which strong correlations exist between two candidate CDEs, we can keep one of the two CDE candidates. Similarly, in the case of attribute-type data element variables, dependencies can be established through the measure of association. In this case, CDE5 and CDE8 were of an attribute type, so an association analysis was performed to determine the relationship between these two potential CDEs. It was found that there was no relationship between CDE5 and CDE8, leading to the decision to keep both of them for DQ assessment.

For the remaining 19 CDEs, a correlation analysis was performed. In the correlation analysis (Table 6.16), there were eight instances (pairs) of strong correlations with correlation coefficients greater than 0.85. These pairs were: CDE4–CDE7, CDE4–CDE10, CDE4–CDE12, CDE10–CDE7, CDE10–CDE12, CDE15–CDE16, CDE13–CDE33, and CDE28–CDE29. In the final step of the prioritization process, it was required that which of the candidate CDEs in a correlation should be kept for DQ assessment be determined. This decision was made through the calculation of S/N ratios.

TABLE 6.16
Correlation Analysis Sample Results

	CDE4	CDE6	CDE7	CDE9	CDE10	CDE12	CDE13	CDE15	…	CDE24	CDE27	CDE28	CDE29	CDE30	CDE33
CDE4	1	–0.158	0.941	–0.070	1.000	0.867	–0.002	–0.048	…	–0.027	0.121	0.057	0.331	0.105	–0.002
CDE6	–0.158	1	–0.136	0.094	–0.158	–0.157	–0.171	0.034	…	–0.012	0.277	0.050	–0.041	0.276	–0.171
CDE7	0.941	–0.136	1	–0.070	0.941	0.822	–0.025	–0.044	…	–0.025	0.128	0.044	0.297	0.187	–0.025
CDE9	–0.070	0.094	–0.070	1	–0.070	–0.066	–0.039	0.015	…	–0.012	0.077	–0.016	–0.050	–0.034	–0.039
CDE10	1.000	–0.158	0.941	–0.070	1	0.867	–0.002	–0.048	…	–0.027	0.121	0.057	0.331	0.105	–0.002
CDE12	0.867	–0.157	0.822	–0.066	0.867	1	0.039	–0.048	…	–0.018	0.122	0.046	0.303	0.087	0.039
CDE13	–0.002	–0.171	–0.025	–0.039	–0.002	0.039	1	0.02	…	0.03	–0.24	–0.22	–0.17	0.04	1.00
CDE15	–0.048	0.034	–0.044	0.015	–0.048	–0.048	0.02	1	…	0.000	–0.013	–0.031	–0.028	0.020	0.020
…	…	…	…	…	…	…	…	…	…	…	…	…	…	…	…
CDE24	–0.027	–0.012	–0.025	–0.012	–0.027	–0.018	0.03	0.000	…	1	–0.038	–0.033	–0.035	–0.043	0.033
CDE27	0.121	0.277	0.128	0.077	0.121	0.122	–0.24	–0.013	…	–0.03828	1	–0.03218	0.1152	0.1743	–0.23989
CDE28	0.057	0.050	0.044	–0.016	0.057	0.046	–0.22	–0.031	…	–0.03277	–0.032	1	0.867	0.132	–0.218
CDE29	0.331	–0.041	0.297	–0.050	0.331	0.303	–0.17	–0.028	…	–0.03514	0.115	0.867	1	0.106	–0.170
CDE30	0.105	0.276	0.187	–0.034	0.105	0.087	0.04	0.020	…	–0.04288	0.174	0.132	0.106	1	0.044
CDE33	–0.002	–0.171	–0.025	–0.039	–0.002	0.039	1.00	0.020	…	0.03347	–0.240	–0.218	–0.170	0.044	1

CDE: Critical data element.

TABLE 6.17
S/N Ratio Analysis for Highly Correlated Variables

CDE	S/N Ratio (dB units)
CDE 4	6.42
CDE 7	6.27
CDE 10	6.47
CDE 12	6.54
CDE 13	11.96
CDE 15	–18.66
CDE 16	–18.57
CDE 28	–4.65
CDE 29	–6.03
CDE 33	11.96

CDE: Critical data element; S/N: Signal-to-noise.

Signal-to-Noise Ratios

Through S/N ratio analysis, one can choose which CDE in a related pair should be kept for DQ assessment. A higher S/N ratio means lower variability. So, CDEs with low S/N ratios require DQ assessment in the form of monitoring and control. Table 6.17 shows S/N ratios analysis for highly correlated CDEs.

Through a correlation matrix (Table 6.15) we can see that CDE4, CDE7, CDE10, and CDE12 have mutually high correlations. CDE7, with an S/N ratio of 6.27 dB, has the lowest S/N ratio in comparison with the others. For this reason, CDE7 was selected for DQ assessment. Similarly, CDE15 was chosen from the pair CDE15 and CDE16. From the pair, CDE28 and CDE29, CDE29 was selected. Since CDE13 and CDE33 had the same S/N ratio, business reasons were applied for the selection and, accordingly, CDE13 was selected for DQ assessment.

So after the final steps, 15 of the original 35 data elements were selected as the final CDEs and DQ assessment was conducted on them. Figure 6.10 shows the funnel approach of reducing the CDEs from 35 in number to 15.

Table 6.18 shows a summary of the DQ assessment corresponding to the 15 CDEs.

Impact on the Analytics Quality

The methodology helped in reducing the number of data elements by selecting a vital of 15. DQ monitoring and controlling were subsequently carried out on these 15 CDEs. These 15 CDEs are typically used in different risk calculators/models to make important decisions. Assuming that there are two models that work based on these 15 CDEs in an analytical function and if the CDE target scores and model performance levels are as shown in Table 6.19, we can estimate the RQI for this

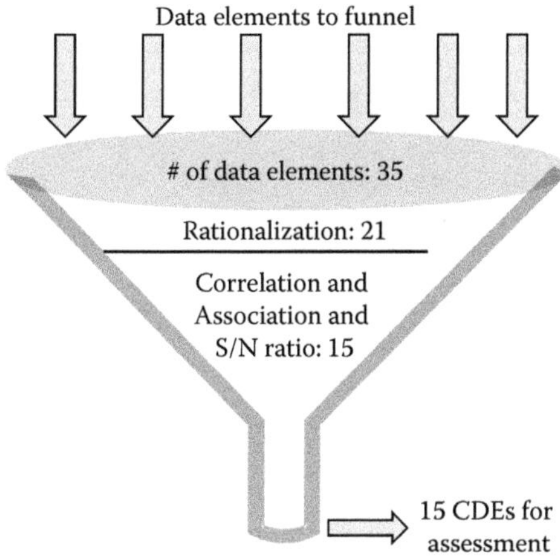

FIGURE 6.10 Funnel approach of reducing the CDEs in number from 35 to 15.

TABLE 6.18
DQ Assessment Summary for the Final 15 CDEs

Dimension CDE	Completeness	Conformity	Validity
CDE5	100.00%	100.00%	100.00%
CDE6	100.00%	100.00%	100.00%
CDE7	100.00%	100.00%	100.00%
CDE8	100.00%	100.00%	100.00%
CDE9	99.80%	99.20%	99.28%
CDE13	0.20%	0.00%	0.00%
CDE15	84.40%	86.60%	86.87%
CDE18	100.00%	100.00%	100.00%
CDE19	100.00%	100.00%	100.00%
CDE20	98.40%	98.00%	98.20%
CDE23	0.60%	1.40%	0.36%
CDE24	100.00%	100.00%	100.00%
CDE27	39.00%	38.00%	32.01%
CDE29	100.00%	100.00%	100.00%
CDE30	100.00%	100.00%	100.00%

CDE: Critical data element.

analytics function. The value of RQI for this analytics function is 22.86. This implies that the overall analytics quality is low, which could be because of the low DQ levels of some of the CDEs and model performance scores. It is important to increase the DQ levels of these CDEs and model performance to make better-quality decisions from the outputs of risk calculators/models.

TABLE 6.19
RQI for Analytics Function (Illustrative Example)

CDE/Model	DQ Score	Target	Squared Deviation
CDE5	100.00	100.00	0
CDE6	100.00	100.00	0
CDE7	100.00	100.00	0
CDE8	100.00	100.00	0
CDE9	99.43	100.00	0.32
CDE13	0.07	85.00	7213.10
CDE15	85.96	95.00	81.72
CDE18	100.00	100.00	0
CDE19	100.00	100.00	0
CDE20	98.20	100.00	3.24
CDE23	0.79	80.00	6274.22
CDE24	100.00	100.00	0.00
CDE27	36.34	90.00	2879.40
CDE29	100.00	100.00	0
CDE30	100.00	100.00	0
Model Performance Score			
Model 1-performance	45	100.00	3025
Model 2-performance	50	100.00	2500
		MSD	34.89
		SNR	−15.42
		Robust quality index (RQI)	22.86

CDE: Critical data element; DQ: Data quality; MSD: Mean square deviation; and SNR: S/N ratio.

6.4 MONITORING AND CONTROLLING DATA QUALITY THROUGH STATISTICAL PROCESS CONTROL TO ACHIEVE ROBUST QUALITY[4]

This case study focuses on applying statistical process control (SPC) in monitoring DQ levels of eligibility data that are related to processing claims in a health care organization. Eligibility data are composed of demographic information for each customer who is eligible to receive medical, pharmacy, dental, or other insurance coverage and indicate if a person can receive benefits with a defined start and end date at a point in time. In the United States health care system, employers (clients) purchase most of the

[4] The author worked with L. Sebastian-Coleman, C. Heien, R. Vadlamudi, and D. Gray on this effort and sincerely thanks them. Special thanks to L. Sebastian-Coleman for providing materials from her publication.

typical benefits for their employees (customers). Since most clients have a relatively stable number of employees, the number of customers that are eligible to receive the benefits (called eligibility counts) is also expected to be stable from month-to-month, with perhaps an exception during the enrollment and renewal periods, at which points employees have the option to change their benefits. The purpose of using SPC on eligibility counts is to proactively determine abnormalities in eligibility data and identify associated data issues. Developing DQ controls directly into applications helps in the quick identification and resolution of data issues.

Analysis Details

Counts of eligible people can be gathered at different levels (e.g., segments, accounts, and products) because they are expected to be consistent at each of these levels. Here, a segment represents the highest level and corresponds to a grouping of clients in different locations. Accounts represent different clients, and products correspond to coverage and programs for sets of customers within a client organization. An example of a product could be a *healthy life program*, wherein health coaches from an insurance company can advise customers on items like food habits and exercise benefits. Measurements at the product level within a client are very granular; measuring them produces large volumes of measurement data. So, this is a nested type of design, as shown in Figure 6.11. Since eligibility data flows are complex, SPC-based controls were applied to segments, accounts within segments, and products within accounts. Measurements at each level help to identify problems at that level. These measurements also help us to understand the impact of downstream problems at higher levels (such as the client- or segment-levels), which makes it easier to isolate and resolve issues. Figure 6.11 provides a generalized, nested view with n segments, k accounts in segments, and m products in accounts. In our analysis, we had only three segments and a specific number of accounts and products for each segment.

Here, we present the SPC analysis discussion for one sample account and one sample product within that account. Figures 6.12 and 6.13 show examples of I-charts

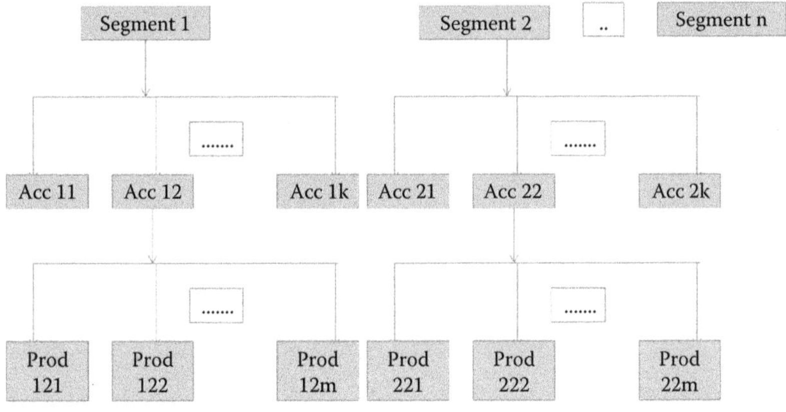

FIGURE 6.11 Nested design for segments, accounts, and products.

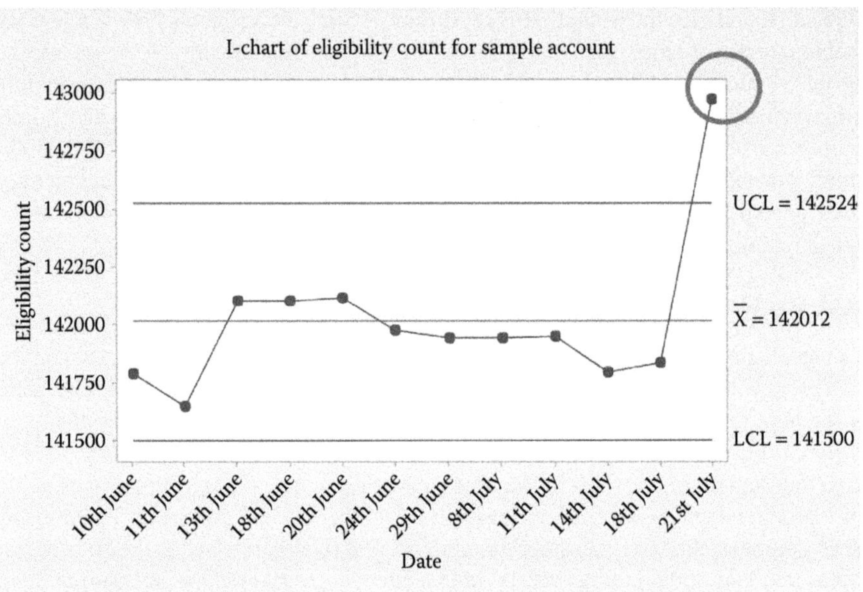

FIGURE 6.12 Thresholds and SPC analysis for eligibility count per account.

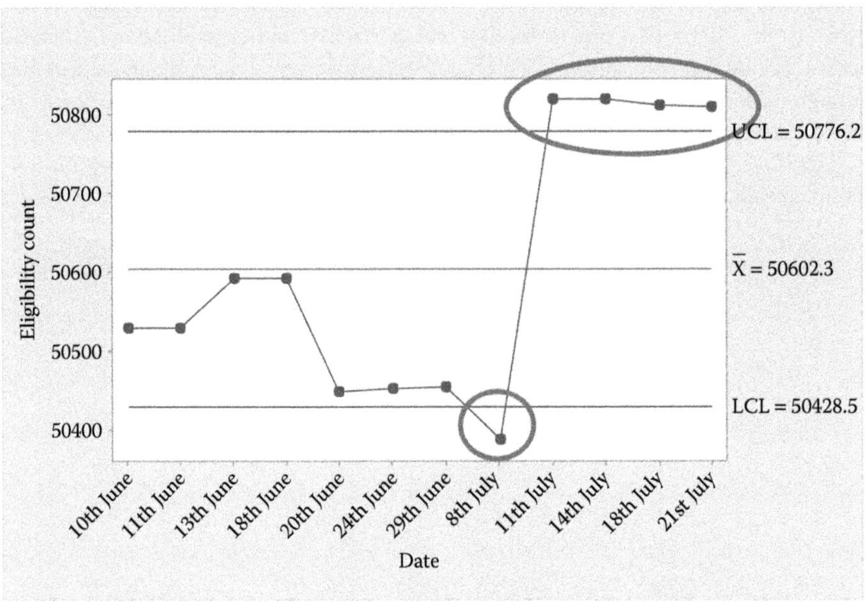

FIGURE 6.13 Thresholds and SPC analysis for counts per day per product group.

(individual control charts) for this account and the product. Since the metric here is eligibility counts, an I-chart is useful, and counts should be between lower and upper control limits of this chart. The upper control limit is located at three standard deviations from the mean on the positive side and the lower control limit is located at three standard deviations from the mean on the negative side. These limits serve as thresholds or specifications and, for an I-chart, they are calculated from moving ranges. Here, a moving range is the absolute difference between two successive observations. The equations to calculate the upper and lower control limits are as follows:

$$\text{Upper Control Limit: UCL} = \bar{X} + 3 \times \left(\frac{\overline{MR}}{1.128} \right)$$

$$\text{Lower Control Limit: LCL} = \bar{X} - 3 \times \left(\frac{\overline{MR}}{1.128} \right)$$

where:

$\bar{X}$ is the mean of daily eligibility counts
$\overline{MR}$ is the mean of all moving ranges.

Appendix I provides procedures to construct different types of control charts with calculations.

Out-of-Control Situations

Figure 6.12 shows an overall increase in the count at the account level and Figure 6.13 shows an increase for a particular product. The eligibility counts are monitored against control limits or thresholds. As mentioned earlier, discrepancies in the counts are expected during the periods of enrollment and renewal, which typically happens at the beginning of the calendar year, while changes in other periods are generally not acceptable. Since discrepancies happened in July, it was required that the issue be investigated by identifying the root cause.

Root Cause Identification and Remediation

These out-of-control situations were discussed with all relevant individuals and the reason for the out-of-control situations was identified: a technical problem in the logic of the program at the account level and product level was identified as the root cause. This problem was resulting in incorrect effective and termination dates. The logic of the program was changed to ensure that the dates were now accurate and the counts were correct. This change brought the counts back into the statistical control.

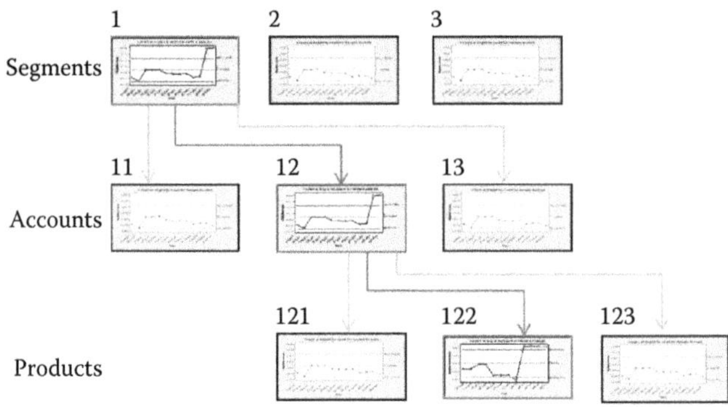

FIGURE 6.14 Tracing root causes through the drill-down of SPC charts to the product level.

Control charts help detect out-of-control situations, as explained earlier. They also help in cases where downstream out-of-control situations impact upstream levels (meaning product level out-of-control situations can have an impact at the account level, and account level out-of-control situations can have an impact at the segment level). Use of an SPC-based control mechanism can help to identify the root causes by drilling down to the product level. This is illustrated in Figure 6.14.

Impact on Process Quality

If the eligibility counts change significantly at other times, it is likely that something is wrong with the business processes. Incorrect counts have an adverse impact on interactions between the insurance company and its customers and clients. Information related to claims payments depends on having complete and reliable eligibility data. So, incorrect or incomplete eligibility data can adversely affect the corresponding processes and can result in customer dissatisfaction, leading the company to lose credibility.

6.5 ANALYSIS OF CARE GAPS[5]

This study was related to gaps in health care. The objective of this study was to understand the drivers of care gaps for a particular disease for health care customers so that suitable actions could be taken to reduce the care gaps. An

[5] The author would like to thank Michael Monocchia for helping in this effort.

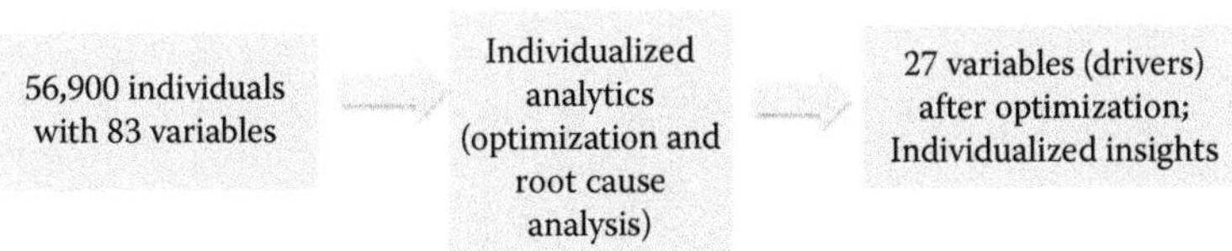

FIGURE 6.15 Individualized analysis of care gaps.

individualized analytics approach was conducted to understand drivers impacting gaps for all of the customers in the study. Figure 6.15 shows high level steps in this approach. A brief description of this approach is as follows:

1. DQ assessment (including dimensions completeness, accuracy, conformity, and validity) was performed on all variables. Through DQ assessment, satisfactory DQ levels were ensured.
2. An individualized analysis was performed involving about 11,700 customers who have received good care (target customers) and about 45,200 customers who have not received good care based on about 83 variables. Based on the results, predictive analytics were performed on other customers (to predict if they will have gaps in their care).
3. The number of variables was reduced to 27 (through system optimization).
4. Root cause driver analysis was completed for all customers with care gaps.

The gaps in care were evaluated by employing distances using multiple variables. Sample individualized results for individuals (denoted as A1, A2, and so on) that have gaps in care are shown in Table 6.20, where highlighted cells indicate a higher percentage of contributions to the distance.

With the results in Table 6.20, the following possible actions can be taken:

- Variables X16, X20, X26, X35, X37, X44, and X48 might be excluded from the programmatic focus as their contributions are quite low to the distances.
- Programs might be built to focus on variables like X2, X3, X27, and X47, as these are high-impact variables.
- Customized programs can be built for the individuals impacted by a particular variable combination.

This type of analysis was effective in understanding the drivers of care gaps on an individual basis by optimizing the information required. This will help in the design of suitable programs/strategies centered around customers so as to reduce gaps in the care.

TABLE 6.20
Sample Individualized Results

	X2	X3	X5	X14	X16	X17	X20	X22	X24	X25	X26	X27	X28	X32	X33	X35	X36	X37	X40	X41	X42	X43	X44	X47	X48	X49	X50
A1	11.16	0.56	62.22	0.52	0.69	0.47	0.26	0.47	0.18	0.19	0.50	1.66	0.03	0.09	0.18	0.55	0.02	0.65	0.45	0.09	0.59	0.55	1.55	0.11	0.00	13.80	0.01
A2	0.01	10.00	7.07	0.25	0.28	0.29	0.32	0.41	0.10	0.27	0.42	28.47	0.35	17.61	0.21	0.39	0.19	0.36	0.02	0.06	0.31	0.56	0.34	27.97	0.83	0.49	0.51
A3	0.48	0.70	0.27	0.11	0.47	0.11	0.12	0.01	0.05	0.04	0.07	0.31	0.18	0.02	0.01	0.16	0.03	0.04	94.04	0.29	0.87	0.44	0.21	0.05	0.04	0.01	0.09
A4	39.62	4.68	0.47	0.24	0.40	0.23	0.27	0.29	0.28	0.09	0.00	16.02	1.88	8.28	1.27	0.25	0.29	0.29	0.03	0.49	0.18	0.24	1.80	12.06	0.01	8.56	0.22
A5	0.03	0.00	0.07	0.62	0.03	0.07	0.00	96.55	0.06	0.04	0.04	0.03	0.05	0.11	0.07	0.10	0.04	0.06	0.05	0.03	0.05	0.10	0.08	0.11	0.07	0.04	0.04
A6	0.59	0.01	0.16	1.43	0.15	0.79	9.48	2.96	34.66	0.12	4.67	0.27	0.00	2.36	8.04	0.21	10.13	0.97	13.19	0.00	2.19	0.01	1.90	0.03	0.02	1.00	0.11
A7	0.01	0.00	0.21	0.10	0.09	0.08	0.08	0.08	0.04	0.01	0.04	0.32	96.90	0.05	0.27	0.07	0.05	0.08	0.07	0.05	0.05	0.08	0.12	0.00	0.16	0.39	0.03
A8	0.86	10.44	5.76	0.27	0.32	0.32	0.33	0.27	0.03	0.35	0.33	28.73	0.27	16.88	0.24	0.46	0.33	0.43	0.00	0.03	0.45	0.35	0.28	27.92	1.03	0.32	0.59
A9	0.16	0.11	0.85	0.16	2.04	0.14	3.69	0.06	0.12	0.01	0.01	0.15	0.24	0.46	0.12	0.13	0.19	0.17	0.11	0.06	0.07	85.47	1.75	0.04	0.07	0.09	0.14
A10	11.51	9.71	15.08	0.29	0.09	0.83	0.54	0.04	0.01	16.47	0.21	0.41	0.10	1.45	0.01	0.50	0.30	0.31	0.42	0.56	0.40	0.72	0.51	0.59	0.44	0.17	12.76
A11	3.40	7.20	0.40	1.32	0.00	0.93	0.44	0.29	1.04	0.38	0.36	5.40	2.03	0.01	0.51	0.63	0.15	1.06	15.85	23.93	1.04	26.13	0.37	0.65	0.51	0.90	1.73
A12	87.52	0.10	3.75	0.12	0.19	0.08	0.22	0.22	0.27	0.03	0.09	0.77	0.27	0.00	0.23	0.23	0.17	0.20	0.00	0.39	0.09	0.18	0.80	0.08	2.17	0.85	0.16
A13	4.17	34.33	44.79	0.42	0.50	0.32	0.02	0.15	0.05	0.17	0.24	1.26	0.15	0.00	0.01	0.37	0.07	0.21	0.04	0.31	0.29	0.53	1.15	0.53	0.00	7.59	0.00
A14	0.43	0.66	2.19	0.24	0.21	0.12	0.07	0.03	0.43	0.08	0.20	0.17	0.16	0.71	90.68	0.14	0.14	0.20	0.01	0.14	0.10	0.16	0.08	0.21	0.00	0.37	0.03
A15	0.02	0.01	0.04	0.08	1.84	83.18	0.17	0.25	0.04	0.15	0.68	2.54	0.64	0.19	0.18	0.13	3.77	0.18	0.17	0.98	0.06	0.03	0.08	0.10	3.08	0.01	0.11
A16	22.96	10.63	10.76	0.82	0.43	0.27	0.25	0.50	0.33	0.23	0.49	1.17	0.35	0.04	0.05	0.36	0.44	0.55	1.20	1.23	32.90	3.75	0.83	0.50	0.02	5.34	0.05
A17	0.00	0.14	0.77	0.03	0.60	0.29	0.24	0.03	0.06	0.21	0.05	0.52	0.06	0.11	0.00	0.20	0.02	0.05	93.82	0.00	0.51	0.66	0.06	0.27	0.28	0.14	0.13
A18	14.77	0.04	0.03	0.60	0.68	0.39	0.65	0.04	0.17	0.73	0.00	25.53	0.07	15.65	1.25	0.29	0.72	0.18	0.04	0.27	0.02	0.06	1.37	29.43	0.44	0.03	0.51
A19	31.24	10.41	0.97	0.51	0.93	0.62	0.24	0.62	0.32	0.05	0.21	3.98	0.00	0.53	0.06	0.31	0.13	0.79	0.32	0.18	0.20	0.28	0.86	0.20	0.01	41.90	0.13
A20	0.37	2.48	0.05	0.15	0.05	0.49	0.64	0.08	0.00	30.34	0.03	0.10	0.00	1.12	0.07	0.23	0.01	0.09	0.32	0.00	0.13	0.54	0.45	0.25	0.11	0.00	25.47
A21	0.02	0.00	4.87	0.01	0.25	0.01	0.10	0.19	0.09	0.02	0.33	0.21	0.01	0.06	0.14	0.45	90.43	0.34	0.01	0.04	0.22	0.23	0.21	0.13	0.07	0.04	0.05
A22	0.12	5.47	0.01	0.13	0.13	0.12	0.07	0.13	0.13	0.76	0.11	15.33	24.53	35.85	0.11	0.11	0.20	0.12	0.15	0.13	0.10	0.14	0.14	12.23	0.20	2.29	0.00
A23	0.53	1.11	0.20	0.15	0.09	0.29	0.00	0.01	0.94	2.98	0.28	0.36	1.15	0.04	72.87	0.49	0.15	0.11	0.00	0.13	0.08	0.20	0.02	0.17	2.28	3.41	1.96
A24	5.65	0.05	2.55	56.84	0.00	2.67	0.02	0.46	0.77	0.75	0.12	1.36	4.98	0.29	0.06	0.05	1.37	0.26	3.16	2.36	0.27	0.41	0.28	2.01	6.47	2.72	1.10
A25	0.12	0.01	0.06	0.14	0.59	0.27	9.80	0.00	0.03	0.07	0.09	0.02	82.72	0.26	0.21	0.22	0.12	0.01	0.02	0.06	2.48	0.26	0.24	0.38	0.14	0.07	0.00

TABLE 6.21
Analytics RQI Care Gap Analysis

DQ/Model	Scores	Target	Squared Deviation
Overall DQ score	96	98	4
Model output	97.5	100	6.25
		MSD	15.12
		SNR	–3.55
		RQI for Analytics	82.25

DQ: Data quality; RQI: Robust quality index; MSD: Mean square deviation; and SNR: S/N ratio.

Calculating Analytics Robust Quality Index

If the overall DQ score was set at 96% against a 98% target and model performance was 98 against the target 100, we can estimate RQI for this individualized analytics model. The value of RQI for this analytics function, as shown in Table 6.21, is 82.25, which can be considered as a satisfactory value.

Appendix I: Control Chart Equations and Selection Approach

ATTRIBUTE CONTROL CHARTS

P-CHART

A p-chart, known as the proportion defective chart, is useful to control and monitor the proportion of defectives. The control limits for the p-chart can be obtained as follows:

Upper control limit: $UCL = \overline{p} + A_0$
Central line = Avg. = $\overline{p}$
Lower control limit: $LCL = \overline{p} - A_0$

Here, $A_0 = 3\sqrt{\dfrac{\overline{p}(1-\overline{p})}{n}}$, which is three times the standard deviation.

NP-CHART

An np-chart is useful to monitor and control the number of defectives (np). The equations to compute the control limits are given in the following:

Upper control limit: $UCL = n\overline{p} + A_0$
Central line = Avg. = $n\overline{p}$
Lower control limit: $LCL = n\overline{p} - A_0$,

where $A_0 = 3\sqrt{n\overline{p}(1-\overline{p})}$, which is three times the standard deviation.

c-Chart

A c-chart, known as a number of defects chart, is used to control the defects instead of the defective units. The requirement in this case is that the inspection unit must be the same for each sample. The control limits for c-chart can be obtained as follows:

Upper control limit: $UCL = \overline{c} + A_0$
Central line = Avg. $= \overline{c}$
Lower control limit: $LCL = \overline{c} - A_0$,

where $A_0 = 3\sqrt{\overline{c}}$, which is three times the standard deviation.

u-Chart

The u-chart is used when we want to monitor and control the average number of defects per inspection unit. If c is the total number of defects in a sample, then the average number of defects u per inspection unit is obtained as $u = c/n$. The following equations are used to construct the u-chart:

Upper control limit: $UCL = \overline{u} + A_0$
Central line = Avg. $= UCL = \overline{u} + A_0$
Lower control limit: $LCL = \overline{u} - A_0$,

where $A_0 = 3\sqrt{\frac{\overline{u}}{n}}$, which is three times the standard deviation.

VARIABLE CONTROL CHARTS

Variable control charts are also known as "twin charts." In these charts, both the process characteristic or variable and its variation are monitored and controlled.

INDIVIDUALS-MOVING RANGE CHARTS

The X-MR chart is a combination of the individual observation (X) and moving range (MR) charts. In an X-MR chart, we first calculate the control limits for the MR chart. To calculate the control limits for the MR chart, the following set of equations is used.

Upper control limit: $UCL = D_4\overline{MR}$
Central line = Avg. $= \overline{MR}$
Lower control limit: $LCL = D_3\overline{MR}$

When the ranges are in control, we can calculate the control limits for the X chart as shown in the following:

Upper control limit: $UCL = \bar{X} + 3\left(\frac{\overline{MR}}{d_2}\right)$

Central line = Avg. = $\bar{X}$

Lower control limit: $LCL = \bar{X} - 3\left(\frac{\overline{MR}}{d_2}\right)$

Note: The values of D_3, D_4, and d_2 depend on the sample size and are given in the following table.

Sample Size (n)	A_2	A_3	B_3	B_4	D_3	D_4	d_2
2	1.880	2.659	0.000	3.267	0.000	3.267	1.128
3	1.023	1.954	0.000	2.568	0.000	2.575	1.693
4	0.729	1.628	0.000	2.266	0.000	2.282	2.059
5	0.577	1.427	0.000	2.089	0.000	2.115	2.326
6	0.483	1.287	0.030	1.970	0.000	2.004	2.534
7	0.419	1.182	0.118	1.882	0.076	1.924	2.704
8	0.373	1.099	0.185	1.815	0.136	1.864	2.847
9	0.337	1.032	0.239	1.761	0.184	1.816	2.970
10	0.308	0.975	0.284	1.716	0.223	1.777	3.078

AVERAGE-RANGE CHART

The average-range chart, or the $\bar{x}$ and R chart, is the most commonly used chart in situations wherein we need to observe drifts in process performance between time periods. Here, also, we first construct a range chart and then the average chart. The following set of equations is used to calculate the control limits.

Range Chart

Upper control limit: $UCL = D_4\bar{R}$
Central line = Avg. of averages = $\bar{R}$
Lower control limit: $LCL = D_3\bar{R}$

When all the ranges are in control, we can calculate the control limits for the averages as shown in the following.

Average Chart

Upper control limit: $UCL = \bar{\bar{X}} + A_2\bar{R}$
Central line = Avg. = $\bar{\bar{X}}$
Lower control limit: $LCL = \bar{\bar{X}} - A_2\bar{R}$

Note: The values of A_2, D_3 and D_4 depend on the sample size and are given in the previous table.

AVERAGE-STANDARD DEVIATION CHART

If the sample size is greater than 9, it is recommended that we use an average-standard deviation chart (or $\bar{x}$-S chart), since the standard deviation is a better estimate of variation for large samples. The procedure for the calculation of control limits for an $\bar{x}$-S chart is similar to that for the $\bar{x}$-R chart. The equations for this chart are given as follows.

S-Chart

Upper control limit: $UCL = B_4\bar{S}$
Central line = Avg. = $\bar{S}$
Lower control limit: $LCL = B_3\bar{S}$

Average Chart

Upper control limit: $UCL = \bar{\bar{X}} + A_3\bar{S}$
Central line = Avg. of averages = $\bar{\bar{X}}$
Lower control limit: $LCL = \bar{\bar{X}} - A_3\bar{S}$

All of the constants in these equations are also given in the previous table. Figure A1 provides a set of approaches to select a suitable control chart. Note that the selection of control chart depends on the type of variable and sample size.

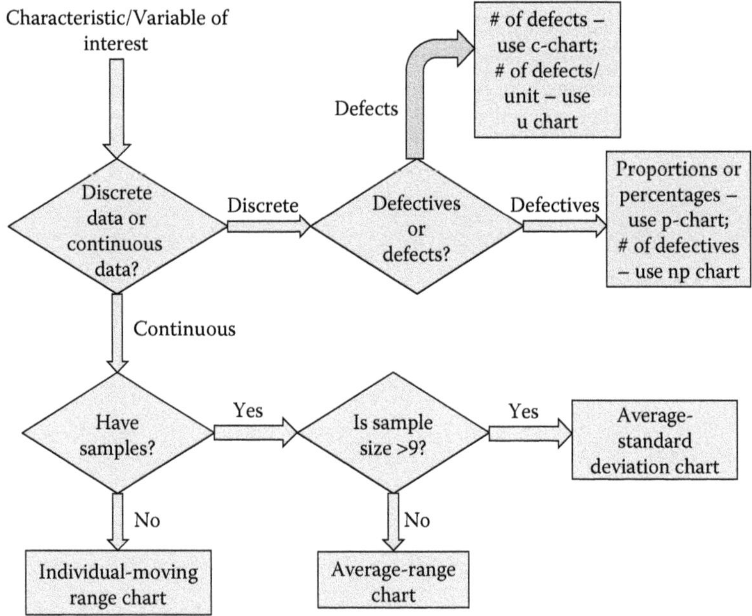

FIGURE A1 Approach to select control charts.

Appendix II: Orthogonal Arrays

EXAMPLE OF ORTHOGONAL ARRAY

Orthogonal arrays are denoted by $L_a(b^c)$
L: Latin square
a: Number of runs or experiments
b: Number of levels of variables
c: Number of columns in the array

EXAMPLE OF TWO-LEVEL ORTHOGONAL ARRAY

$L_{32}(2^{31})$ Orthogonal Array

No.	1	2	3	4	5	6	7	8	9	10	11	12	13	14	15	16	17	18	19	20	21	22	23	24	25	26	27	28	29	30	31
1	1	1	1	1	1	1	1	1	1	1	1	1	1	1	1	1	1	1	1	1	1	1	1	1	1	1	1	1	1	1	1
2	1	1	1	1	1	1	1	1	1	1	1	1	1	1	1	2	2	2	2	2	2	2	2	2	2	2	2	2	2	2	2
3	1	1	1	1	1	1	1	2	2	2	2	2	2	2	2	1	1	1	1	1	1	1	1	2	2	2	2	2	2	2	2
4	1	1	1	1	1	1	1	2	2	2	2	2	2	2	2	2	2	2	2	2	2	2	2	1	1	1	1	1	1	1	1
5	1	1	1	2	2	2	2	1	1	1	1	2	2	2	2	1	1	1	1	2	2	2	2	1	1	1	1	2	2	2	2
6	1	1	1	2	2	2	2	1	1	1	1	2	2	2	2	2	2	2	2	1	1	1	1	2	2	2	2	1	1	1	1
7	1	1	1	2	2	2	2	2	2	2	2	1	1	1	1	1	1	1	1	2	2	2	2	2	2	2	2	1	1	1	1
8	1	1	1	2	2	2	2	2	2	2	2	1	1	1	1	2	2	2	2	1	1	1	1	1	1	1	1	2	2	2	2
9	1	2	2	1	1	2	2	1	1	2	2	1	1	2	2	1	1	2	2	1	1	2	2	1	1	2	2	1	1	2	2
10	1	2	2	1	1	2	2	1	1	2	2	1	1	2	2	2	2	1	1	2	2	1	1	2	2	1	1	2	2	1	1
11	1	2	2	1	1	2	2	2	2	1	1	2	2	1	1	1	1	2	2	1	1	2	2	2	2	1	1	2	2	1	1
12	1	2	2	1	1	2	2	2	2	1	1	2	2	1	1	2	2	1	1	2	2	1	1	1	1	2	2	1	1	2	2
13	1	2	2	2	2	1	1	1	1	2	2	2	2	1	1	1	1	2	2	2	2	1	1	1	1	2	2	2	2	1	1
14	1	2	2	2	2	1	1	1	1	2	2	2	2	1	1	2	2	1	1	1	1	2	2	2	2	1	1	1	1	2	2
15	1	2	2	2	2	1	1	2	2	1	1	1	1	2	2	1	1	2	2	2	2	1	1	2	2	1	1	1	1	2	2
16	1	2	2	2	2	1	1	2	2	1	1	1	1	2	2	2	2	1	1	1	1	2	2	1	1	2	2	2	2	1	1
17	2	1	2	1	2	1	2	1	2	1	2	1	2	1	2	1	2	1	2	1	2	1	2	1	2	1	2	1	2	1	2
18	2	1	2	1	1	1	2	1	2	1	2	1	2	1	2	2	1	2	1	2	1	2	1	2	1	2	1	2	1	2	1
19	2	1	2	1	2	1	2	2	1	2	1	2	1	2	1	1	2	1	2	1	2	1	2	2	1	2	1	2	1	2	1
20	2	1	2	1	2	1	2	2	1	2	1	2	1	2	1	2	1	2	1	2	1	2	1	1	2	1	2	1	2	1	2
21	2	1	2	2	1	2	1	1	2	1	2	2	1	2	1	1	2	1	2	2	1	2	1	1	2	1	2	2	1	2	1
22	2	1	2	2	1	2	1	1	2	1	2	2	1	2	1	2	1	2	1	1	2	1	2	2	1	2	1	1	2	1	2
23	2	1	2	2	1	2	1	2	1	2	1	1	2	1	2	1	2	1	2	2	1	2	1	2	1	2	1	1	2	1	2
24	2	1	2	2	1	2	1	2	1	2	1	1	2	1	2	2	1	2	1	1	2	1	2	1	2	1	2	2	1	2	1
25	2	2	1	1	2	2	1	1	2	2	1	1	2	2	1	1	2	2	1	1	2	2	1	1	2	2	1	1	2	2	1
26	2	2	1	1	2	2	1	1	2	2	1	1	2	2	1	2	1	1	2	2	1	1	2	2	1	1	2	2	1	1	2
27	2	2	1	1	2	2	1	2	1	1	2	2	1	1	2	1	2	2	1	1	2	2	1	2	1	1	2	2	1	1	2
28	2	2	1	1	2	2	1	2	1	1	2	2	1	1	2	2	1	1	2	2	1	1	2	1	2	2	1	1	2	2	1
29	2	2	1	2	1	1	2	1	2	2	1	2	1	1	2	1	2	2	1	2	1	1	2	1	2	2	1	2	1	1	2
30	2	2	1	2	1	1	2	1	2	2	1	2	1	1	2	2	1	1	2	1	2	2	1	2	1	1	2	1	2	2	1
31	2	2	1	2	1	1	2	2	1	1	2	1	2	2	1	1	2	2	1	2	1	1	2	2	1	1	2	1	2	2	1
32	2	2	1	2	1	1	2	2	1	1	2	1	2	2	1	2	1	1	2	1	2	2	1	1	2	2	1	2	1	1	2

EXAMPLE OF THREE-LEVEL ORTHOGONAL ARRAY

$L_{27}(3^{13})$ Orthogonal Array

No.	1	2	3	4	5	6	7	8	9	10	11	12	13
1	1	1	1	1	1	1	1	1	1	1	1	1	1
2	1	1	1	1	2	2	2	2	2	2	2	2	2
3	1	1	1	1	3	3	3	3	3	3	3	3	3
4	1	2	2	2	1	1	1	2	2	2	3	3	3
5	1	2	2	2	2	2	2	3	3	3	1	1	1
6	1	2	2	2	3	3	3	1	1	1	2	2	2
7	1	3	3	3	1	1	1	3	3	3	2	2	2
8	1	3	3	3	2	2	2	1	1	1	3	3	3
9	1	3	3	3	3	3	3	2	2	2	1	1	1
10	2	1	2	3	1	2	3	1	2	3	1	2	3
11	2	1	2	3	2	3	1	2	3	1	2	3	1
12	2	1	2	3	3	1	2	3	1	2	3	1	2
13	2	2	3	1	1	2	3	2	3	1	3	1	2
14	2	2	3	1	2	3	1	3	1	2	1	2	3
15	2	2	3	1	3	1	2	1	2	3	2	3	1
16	2	3	1	2	1	2	3	3	1	2	2	3	1
17	2	3	1	2	2	3	1	1	2	3	3	1	2
18	2	3	1	2	3	1	2	2	3	1	1	2	3
19	3	1	3	2	1	3	2	1	3	2	1	3	2
20	3	1	3	2	2	1	3	2	1	3	2	1	3
21	3	1	3	2	3	2	1	3	2	1	3	2	1
22	3	2	1	3	1	3	2	2	1	3	3	2	1
23	3	2	1	3	2	1	3	3	2	1	1	3	2
24	3	2	1	3	3	2	1	1	3	2	2	1	3
25	3	3	2	1	1	3	2	3	2	1	2	1	3
26	3	3	2	1	2	1	3	1	3	2	3	2	1
27	3	3	2	1	3	2	1	2	1	3	1	3	2

Appendix III: Mean Square Deviation (MSD), Signal-to-Noise Ratio (SNR), and Robust Quality Index (RQI)

MSD	SNR	RQI	MSD	SNR	RQI
100	−20	0	82	−19.14	4.31
99.5	−19.98	0.11	81.5	−19.11	4.44
99	−19.96	0.22	81	−19.08	4.58
98.5	−19.93	0.33	80.5	−19.06	4.71
98	−19.91	0.44	80	−19.03	4.85
97.5	−19.89	0.55	79.5	−19.00	4.98
97	−19.87	0.66	79	−18.98	5.12
96.5	−19.85	0.77	78.5	−18.95	5.26
96	−19.82	0.89	78	−18.92	5.40
95.5	−19.80	1.00	77.5	−18.89	5.53
95	−19.78	1.11	77	−18.86	5.68
94.5	−19.75	1.23	76.5	−18.84	5.82
94	−19.73	1.34	76	−18.81	5.96
93.5	−19.71	1.46	75.5	−18.78	6.10
93	−19.68	1.58	75	−18.75	6.25
92.5	−19.66	1.69	74.5	−18.72	6.39
92	−19.64	1.81	74	−18.69	6.54
91.5	−19.61	1.93	73.5	−18.66	6.69
91	−19.59	2.05	73	−18.63	6.83
90.5	−19.57	2.17	72.5	−18.60	6.98
90	−19.54	2.29	72	−18.57	7.13
89.5	−19.52	2.41	71.5	−18.54	7.28
89	−19.49	2.53	71	−18.51	7.44
88.5	−19.47	2.65	70.5	−18.48	7.59
88	−19.44	2.78	70	−18.45	7.75
87.5	−19.42	2.90	69.5	−18.42	7.90
87	−19.40	3.02	69	−18.39	8.06
86.5	−19.37	3.15	68.5	−18.36	8.22
86	−19.34	3.28	68	−18.33	8.37
85.5	−19.32	3.40	67.5	−18.29	8.53

(Continued)

MSD	SNR	RQI		MSD	SNR	RQI
85	−19.29	3.53		67	−18.26	8.70
84.5	−19.27	3.66		66.5	−18.23	8.86
84	−19.24	3.79		66	−18.20	9.02
83.5	−19.22	3.92		65.5	−18.16	9.19
83	−19.19	4.05		65	−18.13	9.35
82.5	−19.16	4.18		64.5	−18.10	9.52
64	−18.06	9.69		46	−16.63	16.86
63.5	−18.03	9.86		45.5	−16.58	17.10
63	−17.99	10.03		45	−16.53	17.34
62.5	−17.96	10.21		44.5	−16.48	17.58
62	−17.92	10.38		44	−16.43	17.83
61.5	−17.89	10.56		43.5	−16.38	18.08
61	−17.85	10.73		43	−16.33	18.33
60.5	−17.82	10.91		42.5	−16.28	18.58
60	−17.78	11.09		42	−16.23	18.84
59.5	−17.75	11.27		41.5	−16.18	19.10
59	−17.71	11.46		41	−16.13	19.36
58.5	−17.67	11.64		40.5	−16.07	19.63
58	−17.63	11.83		40	−16.02	19.90
57.5	−17.60	12.02		39.5	−15.97	20.17
57	−17.56	12.21		39	−15.91	20.45
56.5	−17.52	12.40		38.5	−15.85	20.73
56	−17.48	12.59		38	−15.80	21.01
55.5	−17.44	12.79		37.5	−15.74	21.30
55	−17.40	12.98		37	−15.68	21.59
54.5	−17.36	13.18		36.5	−15.62	21.89
54	−17.32	13.38		36	−15.56	22.18
53.5	−17.28	13.58		35.5	−15.50	22.49
53	−17.24	13.79		35	−15.44	22.80
52.5	−17.20	13.99		34.5	−15.38	23.11
52	−17.16	14.20		34	−15.31	23.43
51.5	−17.12	14.41		33.5	−15.25	23.75
51	−17.08	14.62		33	−15.19	24.07
50.5	−17.03	14.84		32.5	−15.12	24.41
50	−16.99	15.05		32	−15.05	24.74
49.5	−16.95	15.27		31.5	−14.98	25.08
49	−16.90	15.49		31	−14.91	25.43
48.5	−16.86	15.71		30.5	−14.84	25.79
48	−16.81	15.94		30	−14.77	26.14
47.5	−16.77	16.17		29.5	−14.70	26.51
47	−16.72	16.40		29	−14.62	26.88
46.5	−16.67	16.63		28.5	−14.55	27.26

(Continued)

MSD	SNR	RQI	MSD	SNR	RQI
28	−14.47	27.64	10	−10.00	50.00
27.5	−14.39	28.03	9.5	−9.78	51.11
27	−14.31	28.43	9	−9.54	52.29
26.5	−14.23	28.84	8.5	−9.29	53.53
26	−14.15	29.25	8	−9.03	54.85
25.5	−14.07	29.67	7.5	−8.75	56.25
25	−13.98	30.10	7	−8.45	57.75
24.5	−13.89	30.54	6.5	−8.13	59.35
24	−13.80	30.99	6	−7.78	61.09
23.5	−13.71	31.45	5.5	−7.40	62.98
23	−13.62	31.91	5	−6.99	65.05
22.5	−13.52	32.39	4.5	−6.53	67.34
22	−13.42	32.88	4	−6.02	69.90
21.5	−13.32	33.38	3.5	−5.44	72.80
21	−13.22	33.89	3	−4.77	76.14
20.5	−13.12	34.41	2.5	−3.98	80.10
20	−13.01	34.95	2	−3.01	84.95
19.5	−12.90	35.50	1.5	−1.76	91.20
19	−12.79	36.06	1	0.00	100.00
18.5	−12.67	36.64			
18	−12.55	37.24			
17.5	−12.43	37.85			
17	−12.30	38.48			
16.5	−12.17	39.13			
16	−12.04	39.79			
15.5	−11.90	40.48			
15	−11.76	41.20			
14.5	−11.61	41.93			
14	−11.46	42.69			
13.5	−11.30	43.48			
13	−11.14	44.30			
12.5	−10.97	45.15			
12	−10.79	46.04			
11.5	−10.61	46.97			
11	−10.41	47.93			
10.5	−10.21	48.94			

MSD: Mean square deviation; SNR: S/N ratio; RQI: Robust quality index.

References

Abraham, B. (Ed.). 1998. Process optimization through designed experiments: Two case studies. In Chowdhury, A. R., Rajesh, J., and Prasad, G. K. (Eds)., *Quality Improvement Through Statistical Methods*, Birkhauser publication (now Springer), Boston, MA 263–274.

Albright, S. C., Winston, W. L., and Zappe, C. J. 2009. *Data Analysis & Decision Making* (Revised 3ed.), South-Western Cengage Learning, Boston, MA.

Batini, C. and Scannapieca, M. 2006. *Data Quality: Concepts, Methodologies and Techniques*, Springer, New York.

Blake, R. and Mangiameli, P. 2011. The effects and interactions of data quality and problem complexity on classification. *Journal of Data and Information Quality*, 2(2), 1–28.

Chiang, F. and Renee, M. J. 2008. Discovering data quality rules. In *Proceeding of the 34th International Conference on Very Large Data Bases*, Auckland, New Zealand.

Cong, G., Fan, W., Geerts, F., Jia, X., and Ma, S. 2007. Improving data quality: Consistency and accuracy. In *Proceeding of the 33rd International Conference on Very Large Data Bases*, Vienna, Austria.

Coombs, C. F. 1988. *Printed Circuits Handbook*, McGraw-Hill Book, New York.

Davenport, T. H. and Harris, J. G. 2007. *Competing on Analytics—The New Science of Winning*, HBS Press, Boston, MA.

Deming, W. E. 1993. *The New Economics: For Industry, Government and Education*, MIT Press, London, UK.

English, L. P. 2009. *Information Quality Applied-Best Practices for Improving Business Information Processes, and Systems*, Wiley. Hoboken, NJ.

Gartner. 2012. *Big Data Strategy Components: IT Essentials*, Gartner Publication, Chicago, IL.

Harrington, J. 2006. *The Five Pillars of Organizational Excellence*, Quality Digest (August 2006). http://www.qualitydigest.com/aug06/articles/05_article.shtml.

Herzog, T. N., Scheuren, F. J., and Winkler, W. E. 2007. *Data Quality and Record Linkage Techniques*, Springer, New York.

Huang, K., Lee, T., and Wang, R. Y. 1999. *Quality Information and Knowledge*, Prentice Hall, Upper Saddle River, NJ.

Johnson, R. A. and Wichern, D. W. 1992. *Applied Multivariate Statistical Analysis*, Prentice Hall, Upper Saddle River, NJ.

Jugulum, R. 2014. *Competing with High Quality Data: Concepts, Tools, and Techniques for Building a Successful Approach to Data Quality*, Hoboken, NJ.

Jugulum, R. and Samuel, P. 2008. *Design for Lean Six Sigma: A Holistic Approach to Design and Innovation*, Wiley, Hoboken, NJ.

Jugulum, R. and Frey, D. D. 2007. Toward a taxonomy of concept designs for improved robustness. *Journal of Engineering Design*, 18(2), 139–156.

Juran, J. M. and Blanton, G. A. 1999. *Juran's Quality Handbook*, MacGraw-Hill, Washington, DC.

Leitnaker, M. G., Sanders, R. D., and Hild, C. 1996. *The Power of Statistical Thinking-Improving Industrial Processes*, Addison-Wesley, Reading, MA.

Madnick, S. and Zhu, H. 2006. Improving data quality through effective use of data semantics. *Data & Knowledge Engineering*, 59(2), 460–475.

McFadden, F. R. 1993. Six-sigma quality programs. *Quality Progress*, 26(6), 37–42.

Park, S. H. 1996. *Robust Design and Analysis for Quality Engineering*, Chapman & Hall, London, UK.

Phadke, M. S. 1989. *Quality Engineering Using Robust Design*, Englewood Cliffs, Prentice Hall, NJ.

Phadke, M. S. and Taguchi, G. 1987. Selection of quality characteristics and S/N ratios for Robust design, in *Conference Record, GLOBECOM 87 Meeting*, IEEE communication society, Tokyo, Japan, pp. 1002–1007.

Rao, C. R. 1997. *Statistics and Truth: Putting Chance to Work*, World Scientific, Singapore.

Redman, T. C. 1996. *Data Quality for the Information Age*, Artech House, Boston, MA.

Sebastian-Coleman, L. 2013. *Measuring Data Quality for Ongoing Improvement: A Data Quality Assessment Framework*, The Morgan Kaufmann Series on Business Intelligence. Morgan Kaufmann, Burlington, MA.

Sebastian-Coleman, L., Jugulum, R., Heien, C., Vadlamudi, R., and Gray, D. 2016. Statistical process control and its relevance in data quality monitoring and reporting. *IQ International Journal*, 11(1), 10–22.

Shi, C., Jugulum, R., Joyce, H. I., Singh, J., Granese, B., Ramachandran, R., Gray, D., Heien, C. H., and Talburt, J. R. 2015. Improving Financial services data quality—A financial company practice. *International Journal of Lean Six Sigma*, 6(2), 98–110.

Suh, N. P. 2001. *Axiomatic Design: Advances and Applications*, Oxford University Press, New York.

Suh, N. P. 2005. *Complexity: Theory and Applications*, Oxford University Press, New York.

Taguchi, G. 1986. Introduction to Quality Engineering. Asian Productivity Organization, Tokyo, Japan.

Taguchi, G. 1987. *System of Experimental Design*, Vols. 1 and 2, ASI & Quality Resources, White Plains, New York.

Taguchi, G. 1993. *Taguchi on Robust Technology Development*, ASME Press, New York.

Taguchi, G. and Clausing, D. 1990. Robust quality. *Harvard Business Review*, 68, 65–75.

Taguchi, G. and Jugulum, R. 2002. *The Mahalanobis-Taguchi Strategy: A Pattern Technology*, Wiley, Hoboken, NJ.

Taguchi, G., Jugulum, R., and Taguchi, S. 2004. *Computer-based Robust Engineering: Essentials for DFSS*, ASQ Quality Press, Milwaukee, WI.

Talburt, J. 2011. *Entity Resolution and Information Quality*, Morgan Kaufmann (Elsevier), Burlington, MA.

Wang, R. Y. and Strong, D. M. 1996. Beyond accuracy: What data quality means to data consumers. *Journal of Management Information Systems*, 12(4), 5–33.

Western Electric Company. 1956. *Statistical Quality Control Handbook* (1 ed.), Western Electric, Indianapolis, IN.

Womack, J. P. and Jones, D. T. 1996. *Lean Thinking: Banish Waste and Create Wealth in Your Corporation*, Simon & Schuster, New York.

Wu, C. F. J. and Hamada, M. 2000. *Experiments: Planning, Analysis, and Parameter Design Optimization*, Wiley, Hoboken, NJ.

Index

Note: Page numbers followed by f and t refer to figures and tables respectively.

E

F

G

For Product Safety Concerns and Information please contact our EU
representative GPSR@taylorandfrancis.com
Taylor & Francis Verlag GmbH, Kaufingerstraße 24, 80331 München, Germany

www.ingramcontent.com/pod-product-compliance
Ingram Content Group UK Ltd.
Pitfield, Milton Keynes, MK11 3LW, UK
UKHW021826240425
457818UK00006B/89

* 9 7 8 1 4 9 8 7 8 1 6 5 7 *